T0133211

Editors:
M. Kirkilionis, U. Kummer, I. Stoleriu

2nd UniNet Workshop: Data, Networks and Dynamics

Villa Bosch, Heidelberg
July 3-4, 2006

Bibliografische Information der Deutschen Nationalbibliothek

Die Deutsche Nationalbibliothek verzeichnet diese Publikation in der Deutschen Nationalbibliografie; detaillierte bibliografische Daten sind im Internet über http://dnb.d-nb.de abrufbar.

ISBN 3-8325-1308-6
ISBN13 978-3-8325-1308-5

Logos Verlag Berlin
Comeniushof, Gubener Str. 47,
10243 Berlin
Tel.: +49 030 42 85 10 90
Fax: +49 030 42 85 10 92
INTERNET: http://www.logos-verlag.de

Contents

M. Kirkilionis: Unifying Network Theory?   2

M. Kirkilionis: From graph theory towards dynamical network theory   8

L. Sbano: Semi-microscopic modelling for gene and gene-network   14

U. Janus: Analysis of a synthetic genetic oscillator in *E.Coli*   22

M. Domijan: Bistability in Chemical Reaction Networks   31

S. Bergmann: From DNA arrays to modules to models   40

Y. Bilu: Gene Regulatory Networks   48

Y. Bilu: Studying evolutionary constraints on gene expression regulation in silico   58

J. Bouwman: Vertical Genomics   69

U. Kummer: An Overview of Computational Approaches to Metabolic Networks   72

I. Stoleriu: Stability analysis of metabolic networks   77

I. Vida: Gamma oscillations in interneuron networks   87

D. Battaglia: Networks of neurons   91

D. Battaglia: Chaos in networks with delayed local inhibition   98

J. Bascompte: Ecological networks: methods and data   106

J. Saldaña: Overview: Modelling complex food webs   112

J.L. Garcia-Domingo: Food webs as complex adaptive networks   118

R. Amann: Data Acquisition on Network Formation
in Economics   128

B. Moldovanu: Solution Concepts in Economic Network Formation Models
els   132

T. Gall: Two Applications of Social Networks in Economics   139

Introduction

Introduction: Unifying Network Theory?

Markus Kirkilionis
University of Warwick
Zeeman Building, United Kingdom

Abstract

The UniNet workshop took place in Heidelberg at the Villa Bosch conference center.

The second UniNet workshop was dedicated to the basic scientific idea of the UniNet consortium: to study scientific problems that can be expressed as networks of varying complexity. In all applications UniNet is considering data of different type that can be used to be matched against these different levels of complexity. The most fundamental types of data are related to the network topologies, others related to time series will need to define dynamic processes on the networks.

Background

There has recently much effort to understand complex networks in different areas of applications, predominantly in biology, physics, computer science, sociology and economics. UniNet is covering most of these areas with its different application nodes. The reasons for the current popularity of networks may well be that, like in games popular since centuries like Chess and Go, the concepts and rules of network formation are relatively easy to grasp, but the complexity of possible states and their evolution in time is beyond simple reasoning and therefore interesting to investigate and to compare with real-world data.

Perhaps it is not too far fetched to say that we are currently living in a world of complex networks, at least intellectually. The scientific focus tends to turn away from what we could call mean-field models, i.e. models that describe complex systems with all system components being equal and every component being connected to every other such component. This phenomena is discussed in the following.

Examples of different network applications and data collection

We go through all the physical scales by which the UniNet applications can be ordered. At the molecular level either the binary relation A can bind to B gives rise to an undirected graph, or the functional relation A is regulated by B induces a directed graph, so both types of fundamental network topologies can be found. Examples of corresponding data are protein binding essays for protein-protein (undirected) networks, and micro-arrays for directed network topologies. In both techniques there is not only the network topology to be inferred, but to a limited extend also weights of the resulting graph can be estimated. In the first case these are weights on the links describing affinities, i.e. the strength of the binding, in the case of the micro-arrays these is the strength of the fluorescence signal, i.e. how much response (down- and upregulation) a gene shows with respect to the current environmental conditions of the cell or tissue. Time series are more difficult to measure, but methods like real-time PCR (genetic networks) or NMR and protein kinase phosphorylation (metabolic networks) can deliver valuable dynamical data.

For neuronal networks there are different methods to investigate the synaptic coupling distribution of single neurons establishing the local network topology of a neuronal tissue. Time-series data have been measured since the beginning of the discovery that information processing of neurons is by electric transmission (Hodgkin & Huxley), so is extremely well developed. Data are available from different scales, from single channel recordings to multi-neuron recordings of neuronal electric activity.

In ecology the establishment of graphs describing food-webs is also classical. They are either based on transfer of energy or mass, i.e. the functional relationship are given by A uses energy/mass from B where A, B are trophic levels, populations etc. Time series data in ecology are notoriously difficult to obtain, depending on the time scale of the phenomena. Some indirect effects of whole ecosystem performance can sometimes be obtained, like the whole $CO2$ level in the atmosphere. Algal blooms can be recorded from satellites. But time series from field data having for example a species resolution are very laborious to obtain.

What about the situation in the social sciences and economy? All classical market models are of the mean-field type. The trend is now to identify the local connections that actors and players in the different games structuring an economy (or another such system under scientific observation) are creating. Interestingly what we could call globalisation as a major trend in global economy seems to be (obviously?) described theoretically as having more connections between economic nodes. Does this mean we can go back to the classical market models where due to increased flow of information and capital transactions are completely visible and reachable to the economy as a whole? A warning seems to be in place: Local connections in an interaction network can still be very uneven distributed, and depending on the definition of an economic node local hubs with many more connections than its neighbours are likely to be created. Very often the winner takes all phenomenon is in place, meaning that a node with many links in an evolving network is likely to attract even more links from other nodes, making the link distribution close to a power law. In economic terms such a network is not evolving by poor chance, but

3

active choice and adoption of strategies create preferential links, where the probability of forming a new link is dependent on the so called degree of the node (i.e. the number of links this node already possesses). Corresponding data of such a network formation are now collected, for example using data from the internet, or by questionnaires. Like in ecology dynamical data are harder to collect.

Classes of complexity for networks

Using this more than brief discussion on data collection and its difficulties as a basis it is tempting to think that graph theory (the mathematical theory describing network structures) linked with probability theory would be the major source of mathematical description of complex networks. Indeed such models are currently dominating complex systems theory for good reasons. Nevertheless it is obvious that many biological, ecological, economic and social structures will not be easily mapped to such a framework. Also there is surely a trend and increased technical possibilities to establish more rich time series data sets for different network applications. Taking these facts we can develop corresponding mathematical structures of networks with increased complexity. In the order of low to high complexity the following modelling techniques correspond to the available data:

Static network topologies.

Evolving network topologies.

Static networks with weights attached to either nodes or links.

Evolving networks with weights attached to either nodes or links.

Static networks with temporal changes of weights (i.e. dynamical systems defined on either nodes or links

Evolving networks with temporal changes of weights.

Some explanation seems to be in place. In the discussion on network topologies as above the restriction to modelling will be to the first two types of modelling approaches. We introduced links between proteins, genes, neoron, species and economic players. In all cases there is also the possibility of changing the links over time. But less frequently there is a discussion of the strength of the link. For example models usually become more realistic if the actual affinities, synaptic plasticity or average flow of information between economic or social players would be taken into account. The travelling salesman is another problem example. Here the distance between cities forms a weight on the links of the network without which the problem cannot be properly posed.

Inspired by these examples it is now obvious that the weights defined on the graph are mathematically speaking so-called state spaces. The weights are needed to describe the chemical, biological, ecological, economic or social situation at a given time. The own state of a node (for example the currently available number of molecules, individuals or resources) or of a link (for example again affinities, strength of synaptic coupling, the strength of resource or information flow) is needed to properly map the reality into a theoretical framework.

But now it is obvious that the state on the nodes and links may also vary temporarily. Here we definitely leave the classical area of graph theory and enter a relatively new field, i.e. dynamical systems defined on graphs. There are only a few examples of such models in the literature, most likely due to the difficulties in their investigation. But they are indeed the basic entity of research inside UniNet. Again we can look at the situation with a fixed number of nodes (molecules, individuals, players, economic units) and links, or with a temporarily changing number of nodes and/or links between them. Also both the time and the state space can have either a discrete or a continuous structure. Moreover the dynamics, i.e. the rules with which the state spaces change their state can be either of a deterministic or probabilistic nature.

Examples should shed some light on these concepts. Classical time-continuous game theory with n players, for example with the strategy set defined by the replicator dynamics (Eigen, Schuster, Sigmund, Hofbauer) with applications in biology and economics would constitute a network with n nodes and a time-continuous deterministic dynamical system defined on each of the nodes. The interaction strength between the nodes would be static and given according to a matrix defined at the start of the game. There are obviously many more examples of dynamical systems defined on graphs. Perhaps the most important one for UniNet are reaction networks to which different directed and undirected graph topologies can be associated. The dynamics is again defined on the nodes and are given by concentrations. In this setting important questions of modularity, stability and robustness of dynamic networks can be investigated.

These are all examples of a static network with temporal changes of weights on the nodes. In more general terms biological, ecological, economic and social interaction will nearly always depend on some internal state of the molecules, species, players or agents that can vary in time. Here one of the main tasks of the future is to identify a minimal set of variables for different situations that can adequately reflect such internal dispositions or conformations necessary to describe the experiment. Here mathematics, physics, economy, psychology and social sciences will have to work hand in hand, as they do in UniNet. The same holds for the description of the interactions. In most situations the flow of mass, energy, resources and information between nodes will in turn depend on their own internal states, and vice versa. Again the proper choice of state spaces is the first important and highly non-trivial step. Often the result may well be a multitude of different overlying networks between the nodes describing the different ways in which they interact. Here as always in theoretical approaches there is surely a pay-off between more realistic model representation and mathematical tractability. Most often it is not fruitful to consider too complicated models for which no mathematical results can be obtained. From a philosophy of science point-of-view such models do not exclude any of the possibilities that may constitute the reality around us. In other words they do not enhance our understanding.

5

Conclusion and outlook

The second UniNet workshop will give an overview of the current state of research in the different areas of application UniNet is covering. We will present each node by starting to analyse the available data. We will structure the data according to the UniNet scheme of increasing complexity of networks as outlined above. For this purpose each of the application nodes will invite a guest which should be a specialist in obtaining the corresponding data in this area. This talk will be followed by the theoretical analysis of networks given by UniNet principal investigators.

General Network Theory

From graph theory towards dynamical network theory

Markus Kirkilionis
University of Warwick
Zeeman Building, United Kingdom

Abstract

Networks are prime examples of complex systems. In the simplest case the network topology describes all binary relations between a large number of components. We will call the graph representing the network topology and being associated with the many component complex system the 'interaction graph', with the node set representing the system components one-to-one. From a data acquisition point of view this is at the same time the simplest type of data available, i.e. the data only identify the existence of a component in the system, and whether and with whom it interacts with. A typical complex system research will usually start on such a basis, and if available, will try to incorporate additional data to the network model. Such data are typically ascending in their complexity, quality and richness. The strength of interaction between the components is in many cases measurable, adding or structuring the links or edges of the interaction graph with weights. This can also be interpreted as adding a state space to the links, which in most cases is a real number. It is just a small step to additionally attaching a state space to the nodes, giving the components themselves additional structure. If time is introduced to the system, a law of system evolution can be established. This will either be a mapping of the node state space onto itself (for discrete time), or a stochastic or ordinary differential equation in the case of continuous time.

Introduction

Applying network theory to different complex systems involves several steps which we will outline in the following. The interaction graph is the fundamental entity of any network theory. It is convenient basis from which to start modelling a complex system. It also allows further generalisations, i.e. including more structure by attaching state spaces to nodes and/or edges. This ultimately allows to consider dynamics defined on the networks, with the states changing in time.

Definition of the interaction graph

The interaction graph is given by a directed graph $\vec{G} = (V, E)$, with V being the set of vertices, and E the set of edges, or in this case, arrows. Interactions are interpreted by the relation 'component x is influencing component y', where a component is represented by a single node or vertex. This interpretation gives the edges an orientation. Instead of a symmetric binary relationship such a graph can model a number of other of course identical hierachical relations, like 'x is more dominant than y',, 'x regulates y', etc. The adjacency matrix A of $\vec{G}$ becomes in general non-symmetric. We can define an $(n \times m)$ incidence matrix $D = (d_{ij})$ by $d_{ij} = +1$ if v_i is the positive end of e_j, $d_{ij} = -1$ if v_i is the negative end of e_j, and zero otherwise. The incidence matrix D resulting from giving any arbitrary orientation to G has rank $n - c$, where c is the number of components of G.

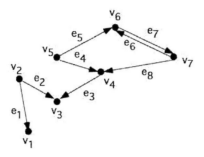

Figure 1: The interaction graph is a directed graph $\vec{G}$

Obviously a binary relation can be modelled as a special case as a pair of parallel arrows pointing in opposite directions. From a modelling point of view the introduction of the interaction graph has several advantages. It fixes the definition of a component, and at the same time defines the possible interaction.

Examples of interaction graphs in different applications

Both the identification of a component and the type of interaction considered can both be trivial or sophisticated from a data point of view. This can nicely be illustrated by the applications considered in this booklet:

Genetic Networks The components forming the set V of the interaction graph are usually the genes, the interaction is given by 'gene x is switching on gene y'. Biologists
often refer to this situation as 'gene y is downstream of gene x'. This example does
illustrate that the definition of a component can be cumbersome when in a later
stage dynamics is introduced. All regulatory elements have to be considered, and
switching might be temporary or even reversed.

Metabolic Networks The components forming the set V of the interaction graph in this case are chemical species, the interaction in the network is given by 'species x is reacting with species y'. In this case the interaction is always symmetric, i.e. the interaction graph could be described as an unweighted graph.

Neuronal Networks The components of the interaction graph are single neurons, the interaction in the network is given by 'neuron x is activating (by synaptic transmission) neuron y'. In this case the interaction graph is (in general) a directed graph $\vec{G}$.

Ecological Networks The components of the interaction graph are single animal or plant species, the interaction in the network is most often given by 'species x is gaining energy (or mass) from species y'. In this case the interaction graph is again a directed graph $\vec{G}$, thought to describe the trophic relations of the food web.

Both static and evolving interaction graphs have been considered in the literature. Evolving graphs have no fixed vertex and edge sets E and V, instead both vertices and edges can be deleted or created by events by given probabilistic rules. This includes mechanisms like preferential attachment, where for example the probability with which a new edge is created between two given vertices in a given event depends on their respective degrees.

Structuring the interaction graph

The next mathematical possibility is to attach state spaces to the (static or evolving) interaction graph, either to the vertices or to the edges. Let X be a finite set (i.e. X is either the set of vertices V or edges E) and $f : X \to K$ with $K = \mathbb{N}, \mathbb{R}, \mathbb{C}$. Usually the set of all such functions f forms a vector space, i.e. the length of roads (think of the traveling salesman problem) usually adds up etc. But this depends on the model. The general idea of introducing state spaces is to attach numerical values to the relationships, or to make these relationships depending on some quantitative measure of the node (like size, richness, status etc.). An obvious generalisation for the set K are vector-valued weights or function state spaces.

After having introduced state spaces rules for updating these states can be introduced. There are several possibilities which should be considered depending on the underlying modelling ideas. For discrete states one can assign transition probabilities making the interaction graph a Markov chain under these assumptions. But the most dominant representative or classes of examples for dynamical network theory in the literature are time continuous dynamical systems defined on the nodes (vertices) of the interaction graph. Consider the n-dimensional system of ordinary differential equations

$$\dot{x}_i = f_i(x_1, \ldots, x_n), \tag{1}$$
$$x_i(0) = x_{i,0}, \tag{2}$$

10

with the flow $\Phi_t(x_0) = (\phi_{t,1}(x_{i,0}), \ldots, \phi_{t,1})$ being the solution of eq (1) - (2). The interaction graph associated to eq. (1) - (2) is given by $V = (v_1, \ldots, v_n)$ (obviously the concentration x_i or mathematically more precisely $\phi_{t,i}(x_{i,0})$ is then associated to vertex v_i), and $e_k = (v_{k_1}, v_{k_2})$ exists if and only if the concentration x_{k_1} is in the list of arguments of f_{k_2}, i.e. the RHS of the k_2th equation of system (1) - (2). We can also interpret the RHS f_i of each equation as dynamically changing weights defined on the edge set E. Note that here we use the notation $(v_i, v_j) \in E$ for an edge or arrow, i.e. we are dealing with ordered pairs of vertices defining the edge. Similarly interaction graphs can be defined for other dynamical systems, like stochastic differential equations, reaction-diffusion systems etc.

This leaves us with the question how the dynamical system defined on the graph structure is derived in different areas of applications. We will deal with this problem in the next sections.

Deriving the network dynamics with multi-scale analysis

Another important aspect of any detailed dynamical network analysis is the derivation of the component's behaviour from microscopic assumptions. This is usually the only way to get a clue what the right dynamics for a problem could be, if one does not like to stick with somewhat more heuristic assumptions. An example for the latter would be for example taking Lotka-Volterra systems as generic equations defined on the network. This has been done quite often historically.

Markov Chains

We can think of a multi-scale analysis in this context as breaking down the components and interactions into micro-structures. This is illustrated in figure 2. The up-scaling consists of taking certain limits, like a fast switching limit where one can average over the different states of the Markov chain over time. This is described in detail in the next chapter by Luca Sbano.

Connection between network topology and qualitative behaviour

Once a dynamical equation has been derived for the interaction graph in an application the question arises whether there is any connection between the network topology (i.e. the graph statistics, like vertex degree distribution, number of simple sub-graphs (or motifs as they are called in Systems Biology) and the qualitative behaviour of the system (in the dynamical system spirit). There is quite complete knowledge about this for simple dynamics, for example for Lotka-Volterra systems (see [1]). To give an illustration, there are theorems ensuring that - for example - the Lotka-Volterra equation is mutualistic, i.e. the Jacobian has only positive off-diagonal entries, any uniquely existing interior equilibrium is a global attractor. Another class of systems, chemical reaction systems

11

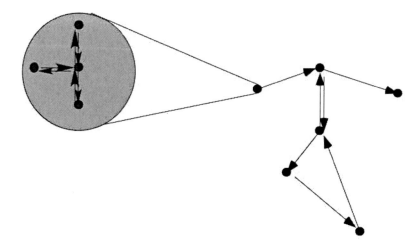

Figure 2: Attaching a Markov chain to a node can be interpreted as first assigning finitely many discrete states to each node, and then defining transition probabilities between these states

described by mass-action dynamics, have rich algebraic structures and a number of results in this direction have been derived. See the different chapters for chemical and metabolic reaction systems in this booklet. For most other systems there are now or very little such precise mathematical results. But numerical investigations can be made. See the chapter on neuronal networks, also especially the chapter on ecological networks in this booklet.

Emergent behaviour of networks

The qualitative behaviour of a dynamical network, i.e. an interaction graph on which a dynamical system is defined, can create what has been described as 'emergent' behaviour of a complex system. The idea is that simple behaviour, or delays of single components create a complex, non-equilibrium qualitative properties of the dynamical network. See here especially the chapter on neural networks where the bifurcation behaviour of a fully connected continuum of neurons creates a rich bifurcation behaviour due to a delay.

Correlation approximation

For networks with discrete states as they are often encountered in epidemiology, or also economics, one can investigate emergent behaviour, i.e. the collective network dynamics, by means of approximations. The idea is to look at the temporal changes of certain configurations in the networks, like the number of linked vertices (pairs of nodes) in which one node is 'infected', the other node is 'uninfected' etc. These so-called correlation equation can exhibit complex behaviour, like oscillations. This also corresponds approximately (depending on the quality of approximation) to the complex emergent properties of the whole network, in this case with discrete states. See for example [2].

Conclusion

Dynamical networks are a challenge for mathematical investigation. They are fundamental for any science of complex systems. Many complex systems can not be analysed by the property of their network topology alone. But they are needed in many applications, so more theory and general properties have to be derived.

Bibliography

[1] Hofbauer, J. and Sigmund, K. *The theory of evolution and dynamical systems*, Cambridge University Press (1988) .

[2] Petermann, T and Paolo De Los Rios. *Cluster approximations for epidemic processes: a systematic description of correlations beyond the pair level*, Journal of Theoretical Biology 229 (2004) p. 111.

Semi-microscopic modelling for gene and gene-networks

Luca Sbano
University of Warwick, Mathematics Institute
Zeeman Building, Coventry CV4 7AL, United Kingdom

Abstract
This project aims to construct an approach to include microscopic effects in modelling genes regulation.

Introduction

Genes activity is usually understood in three phases: *activation, translation* and *transcription.* All these moments involve sequences of molecular processes in which rather different numbers of molecules interact. This can be essentially seen as a problem where many *scales* are involved: *number of particles, time scales* and *spatial scales.* At a macroscopic level these interactions produce the regulation of genes which has rich dynamical aspects. In particular, it has been observed that gene regulation can work at different regimes, this can be understood in an evolutionary context: an organism does not always need to have all its genes expressed at the maximal level, in fact to increase its fitness it is better to have the non necessary ones in a quiescent state ready to switch. The existence of different available useful states is mathematically described by the class of systems called *bistable.* These are dynamical systems with a non-linear dynamics, where the transition between stable states occurs through the action of perturbations. The nature of such perturbations may be both deterministic and stochastic. The natural arising question concerns the way in which the mathematical modelling can approach such systems. At genes level the various different processes not always involve a number of molecules which guarantees the validity of the so called *mass action kinetics.* For example in the *Lac*-operon, the number of repressor molecules is around 10 per cell. In such condition it is clear that one has to consider a microscopic approach to modelling which may allow to take into account effects due the fluctuations.

The theory used is presented in [1], the study of the *Lac-Operon* will appear in a forthcoming paper.

A network perspective

In organisms there are large number of genes and their regulation work through a vast set of rather diverse processes, whose description is often illustrated in terms of graph. The geometry of the graph allows to describe the basic units of the system and their interactions. The main issue is to give a theoretical approach to study these processes from a *systemic* point of view. We shall show that a useful point of view is based on a sequence of up-scaling graphs where at each scale a particular type of graph will describe the relevant processes.

These processes form a complex system, which has a structure common to many other fields of application. Mathematically this combination of the geometry of the graph and the definition can be described using the notion of network where for example the genes and the enzymes form the nodes and the edges represent the reactions. In order to exploit this notion it is necessary to understand the basic processes which constitute the building blocks of the system. We initially concentrate on the study of a single gene case, in order to establish a general and effective approach. In what follows we present how to construct models for genes regulation based on semi-macroscopic point of view. This point of view is based on a description of the basic molecular processes in terms of discrete states and dynamics given by a master equations. The model is then analysed by looking at limits that reproduce a deterministic dynamics together with stochastic correction.

Main result: bistability in the *Lac*-operon

In [8] it is demonstrated how the bi-stable behaviour can be modelled in *Lac*-operon. Here we show that the model presented in [8] can be fully derived and extended by the study of the stochastic model in the regime of adiabatic limit for the genetic activation. A general reference to this approach can be found in [2] and its full construction in [1].

Semi-microscopic model

The model structure we want to present can be defined as *semi-microsopic* because, as it is shown below, reactions are not considered at the quantum level. The steps to construct the model are the following:

- Identify the possible chemical species and divides them into those which are $X_i \in \mathbb{N}$, $i = 1, .., N$ (these may be *large number*) and those $s \in S$, $\dim(S) = M$ that can only undergo to discrete transformations (e.g conformational changes). This produces a state space $\mathbb{X} = \mathbb{N}^N \times S$.

- Set all the possible reactions occurring in the system and interpret the reaction rates as a *probability* rate.

- The system at a certain time t is described microscopically by the joint probability distribution defined on all its possible states $P_s(\mathbf{n}, t) \doteq P_s(n_1, .., n_N, t)$, which solves the so-called *master equation*:

$$\frac{\partial P_s(\mathbf{n}, t)}{\partial t} = L(\mathbf{n}) \circ P_s(\mathbf{n}, t) + K(s, \mathbf{n}) P_s(\mathbf{n}, t), \quad \sum_{s \in S} \sum_{\mathbf{n}} P_s(\mathbf{n}, t) = 1 \quad (1)$$

here L is the generator of the dynamics involving only n_i's degrees of freedom where as the K is the generator of the Markov chain on the finite degrees of freedom s's. In general, K is also function of n_i's.

Multiscale analysis

We study the master equation in the regime were the following two conditions applies:

- large average number $\langle n_i \rangle$ and small fluctuations. To each n_i is associated: $x_i = n_i/\overline{n}$ where $\overline{n} = \max_i \langle n_i \rangle$.

- large frequency transitions in the finite dynamics on S.

These two conditions allow us to construct a multiscale analysis of the process solving the master equation. It is possible to show that when n_i have small fluctuations then the continuous limit is obtained and L becomes a differential operator $\hat{L}$ and the probability $P_s(\mathbf{n})$ becomes a density $\rho_s(x)$. The next step is to consider the large frequency. This corresponds to a system where the evolution on S reaches the equilibrium state faster than the evolution on $\mathbb{N}^N$. In the limit of large number of molecules the master equation can be rewritten as:

$$\frac{1}{\tau} \frac{\partial \rho_s(x, t)}{\partial t} = \frac{1}{\overline{n}} \hat{L}(x) \circ \rho_s(x, t) + k \, K(s, x) \, \rho_s(x, t), \quad \forall s \in S \quad (2)$$

Here τ is the time scaling, $\overline{n}$ is the largest concentration and k is the scaling of the Markov transition probabilities. Since $\overline{n}$ has no dimension we take also τ non-dimensional and equal to $\overline{n}$. Then main assumption is that with $\epsilon = 1/(\overline{n} \, k)$ small. We then consider the time evolution of the marginal probability $f(t, x) \doteq \sum_s \rho_s(t, x)$. As in [2], the time evolution of $f(x, t)$ is given by:

$$\frac{\partial f(t, x)}{\partial t} = (\mathbf{1}^T L) \circ f(t, x) \, \mu - \epsilon \, (L K^+ L)(f(t, x) \, \mu) + O(\epsilon^2) \quad (3)$$

where $\mathbf{1}^T = (1, ..., 1)$ μ is defined by $K\mu = 0$, and $\mathbf{1}^T \mu = 1$, and $K^+ (\mathbf{1} - \mu \mathbf{1}^T) = K^+$. In [1] it is shown that the term $-(\hat{L} K^+ \hat{L}) (\phi(t, \xi) \, \mu)$ is a semi-positive definite parabolic operator, therefore the equation (3) determines a stochastic process whose associated Ito stochastic differential equation has a deterministic drift given by the vector field $\chi(x)$ in the term:

$$\frac{\partial}{\partial x_i} (\chi_i(x) \, f(x, t)) = -\mathbf{1}^T \hat{L}(x) \circ f(x, t) \, \mu,$$

16

and a Wiener term proportional to ϵ. The sde reads:

$$dx_t = \chi(x_t)\,dt + \sigma(\epsilon, x_t)\,dw_t \tag{4}$$

where w_t is the Wiener process, σ is the noise with $\sigma(x, \epsilon) \sim \sqrt{\epsilon}$. To this we have to add the condition that x_t is always in the positive cone $\mathbb{R}_+^N$. It is important to construct the full effective diffusion and then explore the possibility to use stochastic resonance and large deviations method (e.g. Ventze'l and Freidlin theory) to study the bi-stability of the system.

Lac-operon

The *Lac*-operon is one of the most investigated systems of genes (see for example [7] and references therein). The simplest description of the *Lac* can be the following: the operon is induced when the repressor does not bind to the operator sites. When the repressor binds to the operon, LacY is no longer (*on averaged*) expressed. The repressor is tetramer (a dimer of dimers) and it is assumed that can be inactivated by the binding of a single indiucer. LacY is able to uptake from the environment lactose and convert it into an inducer which in turn can bind to the repressor unbinding it from the operon. In the literature (see for example [5]) there is a general agreement on the stoichiometry of the induction. Very recently the analysis carried out in [4] investigated the lac regulation. At the basis of that investigation there is a model that can be described as in figure 1. The state of the operon is give by the state of the main operator sites: Op and Os. This can be described by the 5 symbols:

$$O_{00} = (Op0, Os0),$$
$$O_{10} = (Op1, Os0), \quad O_{01} = (Op0, Os1),$$
$$O_{11} = (Op1, Os1), \quad O_L = Loop$$

0 means free site and 1 binding to the repressor.
It will be convenient to define the set:

$$S = \{(00), (10), (01), (11), (L)\}.$$

The state space

In order to built a model we need to identify the "state space" for the system: r_0 the number of free repressors, primary operator Op and secondary operator Os, x molecules of free inducer TMG, y molecules of LacY, c complex produced during the the uptake of TMG.
The state is:

$$\xi = (x, r_0, y, c, (ij))$$

$x, y, c \in \mathbb{R}$ and $(ij) \in S$.

17

Remark

The state ξ depends also on the number of free available repressors r_0. Free repressors have a dynamics that is not considered in this simplified model, therefore r_0 appears as a parameter.

The objective is to compute the probability distribution:

$$P_{ij}(t, x, r, y, c; r_0).$$

The system has a state that at time t is described by: r_0 free repressors, x free inducers, Op principal operator, Os secondary operator.

Processes

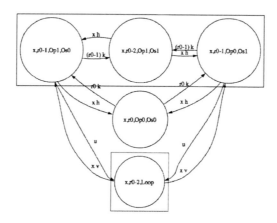

Figure 1: Lac with two operators. In the picture the transition rates are equal for O_s and O_p.

We now describe the main processes forming the whole gene regulation:

Gene activation

$$(x, r_0, Op0, Os0) \underset{x\,h_p}{\overset{r_0\,k_p}{\rightleftharpoons}} (x, r_0 - 1, Op1, Os0), \quad (x, r_0, Op0, Os0) \underset{x\,h_s}{\overset{r_0\,k_s}{\rightleftharpoons}} (x, r_0 - 1, Op0, Os1)$$

$$(x, r_0 - 1, Op1, Os0) \underset{x\,h_s}{\overset{r_0\,k_s}{\rightleftharpoons}} (x, r_0 - 2, Op1, Os1), \quad (x, r_0, Op0, Os1) \underset{x\,h_p}{\overset{r_0\,k_p}{\rightleftharpoons}} (x, r_0 - 2, Op1, Os1)$$

$$(x, r_0 - 2, \text{Loop}) \underset{u_s}{\overset{x\,v_s}{\rightleftharpoons}} (x, r_0 - 1, Op1, Os0), \quad (x, r_0 - 2, Op0, Os1) \underset{u_p}{\overset{r_0\,k_p}{\rightleftharpoons}} (x, r_0 - 2, Op1, Os1)$$

$$(5)$$

which corresponds to reactions during which (x, y, c) are conserved.

18

Expression, up-taking and degradation

- Expression of LacY

$$(Op_i, Os_j) \to^{\gamma_{ij}} LacY, \quad O_{Loop} \to^{\gamma_{Loop}} LacY$$

- up-taking of TMG:

$$LacY + TMG_{ext} \underset{l_2}{\overset{l_1}{\rightleftharpoons}} C \to^m LacY + TMG$$

We shall term $TMG_{ext} = T$.

- Degradation processes:

$$LacY \to^{\delta_y} \emptyset, \quad TMG \to^{\delta_x} \emptyset$$

here TMG_{ext} will be considered constant. The LacY forms a complex C with the Allo-lactose which in turn is up-taken into the cell, there the compound C is degraded into X and Y.

$$Y \to^{\delta_y} \emptyset, \quad X \to^{\delta_x} \emptyset \tag{6}$$

Dimensions

It is useful to give explicitly the the dimensions of the various reaction constants:

- Activation: $[k_i] = [h_i] = 1/(time \cdot \#particle)$, $[u_i] = 1/time$, $[v_i] = 1/(time \cdot \#particle)$ where i is either p or s.

- Translation: $[\gamma_i] = \#particle/time$,

- Up-taking and degradation: $[\delta_x] = [\delta_y] = 1/time$, $[l_1] = 1/(time \cdot (\#particle)^2)$, $[m] = [l_2] = 1/(time \cdot \#particle)$.

Scaling analysis and adiabatic limit

The master equation turns out to to be of the form (1). We now perform a time scaling analysis and obtain that the deterministic dynamics is given by:

Proposition 1
The adiabatic limit equations describing the time evolution of x, y, c are:

$$\begin{cases} \dot{x} = -\delta_x\, x + m\, c \\[2mm] \dot{y} = -\delta_y\, y + (-l_1\, T\, y + (l_2 + m)\, c) + \dfrac{q_2(x, r_0)}{p_2(x, r_0)} \\[2mm] \dot{c} = l_1\, T\, y - (l_2 + m)\, c \end{cases} \tag{7}$$

where $p_2(x, r_0), q_2(x, r_0)$ are two explicit polynomials in x and r_0. If $u_s = u_p$, $v_p = u_p$, $k_s = k_p$ and $h_p = h_s$, then:

$$\frac{q_2(x, r_0)}{p_2(x, r_0)} = \frac{x^2\,\gamma_{00} + K\,r_0\,x\,(\gamma_{10} + \gamma_{01}) + r_0\,(r_0 - 1)\,K^2\,\gamma_{11} + r_0\,K\,U\gamma_{Loop}}{x^2 + K\,r_0\,x + (r_0 - 1)\,K^2 + r_0\,K\,U}$$

where $U = u/v$ and $K = k/h$.

Proof. Let us define:

$$f(t, x, y, c) = \sum_{(ij)\in S} \rho_{ij}(t, x, y, c)$$

we have seen that for large $\bar{n}$ the limit equations for f is given by:

$$\frac{\partial f}{\partial t} = \mathbf{1}^T \hat{L}(\mu\,f)$$

where all terms of order $o(1/\bar{n})$ are neglected. Now we have:

$$\mathbf{1}^T \hat{L}(\mu\,f) = (\sum_{(ij)\in S} \hat{L}_{ij}\mu_{ij})f.$$

which gives, after some calculations (7). **Q.E.D.**

Quasi-steady state approximation

In order to derive the model presented in [5] we need to perform another approximation, namely we assume that *the complex C is at a steady state.* This implies:

$$\dot{c} = 0$$

hence from (7):

$$c = \frac{l_1\,T}{l_2 + m}\,y$$

therefore equations (7) simplifies to:

$$\begin{cases} \dot{x} = -\delta_x\,x + \dfrac{m\,l_1\,T}{l_2 + m}\,y \\[2mm] \dot{y} = -\delta_y\,y + \dfrac{q_2(x, r_0)}{p_2(x, r_0)} \end{cases} \tag{8}$$

One can easily verify (see [5]) the following:

Proposition 2
The system (8) is generically bi-stable.

Future directions

- Analysis of the bi-stability by looking at the stochastic fluctuations (e.g. small occupation numbers effect and noise): large deviation and stochastic resonance.

- Use many-body methods to evaluate fluctuations (see [6])

- Integrate many genes to form genetic networks. Is it emerging a modular structure?

Bibliography

[1] L. Sbano *Notes on stochastic reaction networks* Warwick Preprint 09/2006

[2] T.B. Kepler and T.C. Elston *Stochasticity and Transcriptional Regulation: Origin, Consequences, and Mathematical Representation* Biophysical Journal **81** (2001)

[3] B. Lewin *Genes* VI edition, OUP (1998)

[4] JMG Vilar, L. Saiz *DNA looping in gene regulation: from the assembly of macromolecular complexes to the control of transcriptional noise* Current Opinion in Genetics and Development (2005) **15**

[5] E. M. Ozbudak, Mukund Thattai, Han N. Lim, B. I. Shraiman and Alexander van Oudenaarden *Multistability in the lactose utilization network of Escherichia coli* Nature **427** 2004

[6] M.Sasai and P.Wolynes *Stochastic gene expression as a many body problem* PNAS **100** (2003)

[7] M. Lewis *The lac repressor* C. R. Biologies **328** (2005) 521548

[8] E. M. Ozbudak, M. Thattai, H. N. Lim, B. I. Shraiman and A. van Oudenaarden *Multistability in the lactose utilization network of Escherichia coli* NATURE, 2004

[9] J.M.G. Vilar and S. Leibler *DNA looping and physical constraints on transcription regulation* J.Mol.Biol. 2003, **331**.

Analysis of a synthetic genetic oscillator in *E. Coli*

Ulrich Janus
University of Warwick, Mathematics Institute
Coventry CV4 7AL, United Kingdom

Abstract
Constructions of small synthetic genetic networks have opened up a new perspective in studying the workings of transcriptional regulation. These engineered genetic circuits have the advantage of being isolated from outside influences and made up of well characterized components, which makes them accessible to mathematical modelling. Of particular interest in genetic regulation are the role of noise and certain types of dynamical behaviour like oscillations, which are important for biological time keeping mechanisms. This project focuses on the analysis of a small two gene oscillator implementing an autoactivating and a repressive feedback loop. Guided by dynamical and stochastic modelling, we are aiming tracing the system's transition though a Hopf bifurcation by mutagenesis of the involved promoter regions.

Introduction

Interactions and interdependencies of genes and their expression products form large and complex networks known as genetic regulatory networks. The regulation of the genes enables organisms to change the pattern of gene expression over time and produce proteins in response to interior and exterior stimuli. The network models applicable in this context are those of dynamic nodes and static links. The links correspond to the regulatory interaction strengths governed by binding affinities of transcription factors and repressors, or number and location of operator sites. The nodes are usually identified with the genes and proteins, where the first may in different states of transcription activity, while for the latter the number of molecules is of relevance. This means that genetic regulatory networks are an example of a dynamical system defined on a network. Here time-series data of expression levels of a small gene network in concert with mathematical modeling will be used to study its dynamic behavior, in particular the occurrence of oscillations.

In most organisms a number of genes is expressed rhythmically. This phenomenon is usually referred to as a biological or circadian clock. It is thought that most biological clock are driven by a central oscillator made up of one or several negative genetic feedback loops. The dynamics of these central oscillators are typically difficult to study as they are embedded into complex and often poorly known regulatory network which integrates ‚time-setting' signals such as light and temperature. To address this problem recent research has been focusing on the generation of synthetic genetic circuits made from well understood natural components [7], [9], [11]. This project will focus on the occurrence of oscillations of gene expression levels in a simple two gene network described by Atkinson [3]. This circuit was constructed from components of the Lactose and Nitrogen metabolic subsystems of E.Coli and implements a positive andnegative feedback loop between the two genes, Mutagenesis of interaction sites in the promoter region is expected to enable the manipulation of binding affinities and regulatory efficacy in a controlled manner. The analysis of a mathematical model will identify the systems sensitivity to changes of different parameters. We hope that this setup will allow the traversing of (Hopf) bifurcation points by biological means.

For the mathematical model of gene networks, diverse processes like transcription, translation, degradation and transitions between activity states of genes have to be taken into account. The approach used here is that of the ‚semi-microscopic' modeling described in the previous section by Luca Sbano, where the protein-DNA interactions in the promoter region of a gene are described by stochastic processes and fast operator or diffusion limits are taken to derive (stochastic) ordinary differential equations.

Synthetic Gene Networks

The regulatory interdependencies amongst the genes in a cell form a highlycomplex and dynamic network. The product of any gene may trigger directlyor indirectly the activation or repression of one or more other genes. Theseeffects may happen fast or with a time delay, they may be switchlike ormore gradual. From this variability of regulatory processes and the sheersize of cellular genomes arises the complexity of natural gene networks.

Natural Systems versus Engineered Circuits Recent advances in the technology of microarrays havemade available genomic and proteomic data on a large scale. This data isnow being used to identify certain structural modules within a gene networkand to elucidate the dependency relations between genes [1]. So far thereis not enough information which would make dynamical modeling on thisscale feasible. Instead, mathematical models of natural genetic networkshave focused on interesting subsystems [15], [17]. But even for the smaller systems the data is typically sparse and many important components are not known, which imposes a strong limitation on the reliability of these models. An alternative approach to the study of naturalsystems is the construction and analysis of synthetic gene circuits from wellunderstood natural components [11]. This procedure yields a small and selfcontainedregulatory network, whose interactions with the rest of the cellsgene regulation network are tractable. Due to the isolation from outside influencesthe activities of individual genes can be measured more accuratelythan would be possible in natural systems. Such a network enters the scopeof detailed, dynamical mathematical modeling in close relation to the experimentaldata.

Approaches and Applications: This field of synthetic biology has yielded a range of interesting research opportunities as well as applications in medicine and bioengineering.The typical approach, referred to as forward engineering, is to conceptualise a genetic switch, analyse it by some basic dynamical modelling and then to create the corresponding plasmids and cell lines and analyse the circuits behavior by experiments. One example for this is the Repressilator by Elowitz and Leibler [7], an oscillating three gene network driving by a repressive feedback loop. Other examples include a toggle switch [9] and elementary logical gene circuits inspired by the idea of gene computing [10].

Synthetic circuits have been applied to study dynamical characteristics of gene regulation: Becskeiand Serrano used an engineered circuit, also in E. coli, to investigate the stabilising roleof feedback loops [5].

Also it has become clear that noise plays am important role in gene regulation [22]. Synthetic gene circuits have helped to elucidate the role of noise propagation in gene expression [24], [21] and in probabilistic cell differentiaion [28].Promising medical applications include the analysis of the mode of action of toxins. Engineered RNA based gene regulators allow a controlled turning on of genes and so to study the effects of their products in a controlled manner [12].The counterpoint of the forward engineering approach is inferring the network topology from gene expression data [29].Alternatively, if given the network architecture and gene expression patterns, It might be possible to identify drug targes or the modes of toxin action [8],[6].This project will use the experimental tractability of synthetic gene networks to attempt subtle small scale network manipulation guided by detailed mathematical modelling.

Biological background of the synthetic clock

In the following I give a quick overview of the synthetic clock constructed byAtkinson [3] and describe how its components are regulated.

Design of the gene circuit

The network comprises two modules, an activator and a repressor, whichfeed back into each other. See figure . Each module consists of a gene and a promoter sequence including certain enhancer sites. The modules are constructed from components of the regulatory systems for lactose (Lac operon) and nitrogen metabolism.

The activator module implements and autoregulatory positive feedback loop. The gene *glnG* was fused to the *glnAp2* promoter sequence. The product of the gene *glnG* is the protein NRI. In its phosphorylated form (NRI~P)it activates the glnAp2 promoter. Thus NRI~P activates its own transcription. To receive the signal from the repressor module the *glnAp2* promoter has been modified by inserting two Lac operator sites.

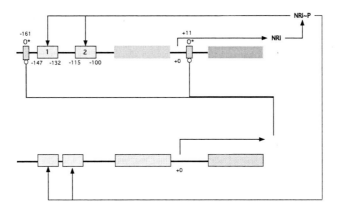

Figure 1: **Diagram of regulatory interaction of the synthetic oscillator.**

The repressor module consists of the *lacI* gene fused to the *glnKp* promoter. Similar to the *glnAp2* promoter, the *glnKp* promoter is activated by NRI~P binding to two adjacent binding sites.

Transcriptional isolation was achieved by inserting the modules within landing pads [3]. The cell carrying the modules had mutations in the chromosomal genes *lacI*, *glnG*, *glnL*, so that they were unable to produce functional LacI, NRI and NRII. NRII was instead provided by a plasmid [3]. So the only source of LacI and NRI was the synthetic clock.

Current understanding of the regulatory interactions

NRI dependent transcription at the glnA promoter: In wild type E.coli the expression of the gene *glnA* is driven by two promoters *glnAp1* and *glnAp2* [23]. The first promoter *glnAp1* is dependent on the catabolite activator protein (CAP) and is repressed by NRI. The second promoter *glnAp2* is dependent on the phosphorylated form of NRI, denoted by NRI~P.

The phosphorylation state of NRI is controlled by the kinase/phosphatase NRII [18]. Unphosphorylated NRI forms inactive dimers. Upon phosphoylation the dimeric form is no longer stable but reassembles to form heptamers, which expose the regulatory domains which interact with σ^{54} [14].

The factor NRI~P initiates transcription by catalyzing the transformation of the closed transcription complex into the open one. This interactions is facilitated by two enhancer sites NRI1 and NRI2 upstream of glnAp2 [19]. The NRI~P bound to the enhancer sites is brought in contact with the σ^{54}-RNA polymerase bound at the promoter by the formation of a DNA loop [27].

At high concentrations of NRI~P the activity of the *glnAp2* promoter is again reduced [26]. Responsible for this effect are three low-affinity NRI~P binding, or governor sites

NRI3, NRI4, NRI5 [4]

The activation of the promoter glnKp works similarly to that of *glnAp2*. Here also two enhancer sites, one with a high, the other with a low affinity for NRI~P facilitate its interaction with the σ^{54}-RNA polymerase. The *glnKp* promoter needs a higher NRI~P concentration for activation [2].

The DNA loop formed by the LacI repressor and the twoperfect operator sites inserted into the glnAp promoter region is thought to interfere with theNRI activation loop. Also LacI bound to the operator site at +1 could block thesuccessful assembly of the transcription complex.

The modeling approach

This section outlines the derivation of a model of the synthetic clock from the fundamental molecular interactions between the regulatory factors and their respective interactive sites, which control transcription of the two genes.

Towards a discrete probabilistic model

At the onset we will formulate the the regulatory mechanisms as discrete states like the binding status of operator and enhancer sites, or the topology of the DNA (loop or no loop), and probabilistic transitions between them. The different states and corresponding transition probabilities will define a Markov process, for which a Master equation, i.e. a differential equation for the probabilities to be in a certain state at a given time, can be derived. The Master equation can become quickly very unwieldy and is typically difficult to solve except for simple cases. Therefor one aims at taking certain limits like diffusion or fast operator fluctuations, which lead to stochastic differential equations. The benefit of this approach is that the noise terms have a clear interpretation, which is sometimes not that straightforward if noise terms are added post-hoc. The stochastic differential equations can then be treated with bifurcation analysis, either deterministic ally or with a generalized probabilistic interpretation of steady states and bifurcation. The ODE analysis may be complemented with computational analysis of the discrete model by MCMC methods like the Gillespie algorithm.

Formulation of the discrete state model

The discrete state approach to genetic regulation has been reviewed by Kepler and Elston [13] illustrated by examples of simple activating factors. Details on the derivation and expansion of the Master equation are described in the book by van Kampen [30]. The repression by LacI is discussed in detail also by Sbano [25]

As an example for the first stage in the development of the model for the synthetic oscillator we present a simple model for the formation of the DNA by the binding of LacI to two operator sites O_1, O_2. As the LacI repressor is a dimer of dimers, which can be present as an assembled tetramer or as dimer, two ways towards the loop formation are possible [16]: The two dimers could bind two the two operators and form a loop by assembling the tetrameric structure. Alternatively, one of the dimers of a full tetramer could bind two one of the operators and the loop is formed then by the interaction of the second dimer with the second operator.

The models arising from these alternative mechanisms and defined by the different states and possible transitions are depicted in figure .

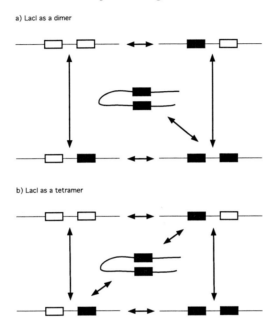

Figure 2: **Binding states of LacI repressor and the operators. a) Dimer model. b) tetramer model.**

We note here the Master equation for the dimer model. The LacI repressor is assumed to be present at a constant amount m. The state of the system is defined by the pair (O, L). The operator states are $O = 0$ for two free operator sites, $O = 1$ for LacI bound to one operator sites (assumed to be symmetric) and $O = 2$ for both operator sites occupied. The loop state L can change between $L = 0$ for no loop and $L = 1$ for the formed loop.

$$
\begin{aligned}
\dot{p}_{00}^m &= -mk_b^+ p_{10}^m \\
\dot{p}_{10}^m &= -(k_b^- + (m-1)k_b^+)p_{10}^m + mk_b^+ p_{00}^m + k_b^+ p_{20}^m \\
\dot{p}_{20}^m &= -(k_b^- + k_l^+)p_{20}^m + (m-1)k_b^+ p_{10}^m + k_l^- p_{21}^m \\
\dot{p}_{21}^m &= -mk_l^- p_{21}^m + k_l^+ p_{20}^m,
\end{aligned}
$$

where k_b, k_l are the rate coefficients for repressor binding and loop formation respectively, and the superscript $\pm$ indicates the direction of the transition. Further work will see the derivation and approximate solution of a complete model for the described synthetic clock.

Relation to deliverables of work package 2

Work on: M1-model formulation for generic genetic networks. Genetic transcription networks can formally be described by reaction schemes like for chemical interaction networks. But the nature of the interactions is somewhat different. For once stoichiometric constraints derived from mass and energy balance does not hold for transcription networks. But more importantly the number of the involved species is typically very low. That means that the stochasticity of the molecular events cannot be neglected like is usually done for chemical reaction networks. So a generic model for genetic networks has to treat the probabilistic nature of these interactions with care. Consequently I am currently working on the derivation of a model for a small transcription network starting from very basic discrete molecular interactions as is described in the project report. This approach would be applicable to general reaction networks, where diffusion limits do not necessarily hold.

Work on: network dynamics and its qualitative behaviour. Dynamical networks show a wide variety of possible qualitative behaviours, like multi-stability, oscillations or even chaos. For many applications oscillatory behaviour is of particular interest. It is known that positive and negative feedback loops are important in the generation of stable oscillations. In the case of genetic regulatory or neural networks it seems that often multiple feedback loops or subnetworks capable of limit cycle dynamics interact to produce the observed oscillations. In this context the synthetic oscillator under study may serve as a model system how oscillation in this kind of networks may arise.

Work on: The reduced network. Naturally occurring networks are typically large and complex, which imposes strong limitations on the sophistication of the dynamics which can be studied. Finding techniques for the reduction of this complexity is hence an ongoing challenge. In particular in chemical reaction networks the simplifying assumption of fast diffusion and averaging out of the fluctuation of the discrete interaction events is usually made to facilitate analysis and computations. In other systems like gene networks is assumptions is not always feasible. Still, the need for simplification persists as probabilistic modelling of discrete random events becomes intractable very quickly. During the course of the synthetic oscillator project techniques will be reviewed how limits of fast operator or diffusive fluctuations can be made and when they are mathematically justified.

Bibliography

[1] Allison, David, B. et al., *Nat Rev Genet*, **7**, 1, 55–65, (2006).

[2] Atkinson, M. R. et al., *J. Bacteriol.*, **184**, 19, 5358–5363, (2002).

[3] Atkinson, M., R. et al., *Cell*, **113**, 5, 597–607, (2003).

[4] Atkinson, M., R., Pattaramanon, N. and Ninfa, A. J., *Molecular Microbiology*, **46**, 5, 1247–1257, (2002).

[5] Becskei, A. and Serrano, L., *Nature*, **405**, 6786, 590–593, (2000).

[6] di Bernardo, D. et al., *Nat Biotech*, **23**, 3, 377–383, (2005).

[7] Elowitz, M. B and Leibler, S., *Nature Nature*, **403**, 6767, 335–338, (2000).

[8] Gardner, T. S., di Bernardo, D., Lorenz, D. and Collins, J. J., *Science*, **301**, 5629, 102–105, (2003).

[9] Gardner, T. S., Cantor, C. R. and Collins, J. J., *Nature Nature*, **403**, 6767, 339–342, (2000).

[10] Guet, C. C., Elowitz, M. B., Hsing, W., and Leibler, S., *Science*, **296**, 5572, 1466–1470, (2002).

[11] Hasty, J.,McMillen, D. and Collins, J. J., *Nature*, **420**, 6912, 224–230, (2002).

[12] Isaacs, F. et al. *Nat Biotech*, **22**, 7, 841–847, (2004).

[13] Kepler, T. B. and Elston, T. C., *Biophys. J.*, **81**, 6, 3116–3136, (2001).

[14] Lee, S.-Y. et al., *Genes Dev.*, **17**, 20, 2552–2563,(2003).

[15] Leloup, J.-C. and Goldbeter, A., *J Biol Rhythms*, **13**, 1, 70–87, (1998).

[16] Lewis, M., *Comptes Rendus Biologies*, **328**, 6, 521–548, (2005).

[17] Locke, J. C. W., Millar, A. J. and Turner, M. S., *Journal of Theoretical Biology*, **234**, 3, 383–393, (2005).

[18] Ninfa, A. J. and Magasanik, B., *PNAS*, **83**, 16, 5909–5913, (1986).

[19] Ninfa, A. J., Reitzer, L. J. and Magasanik, B., *Cell*, **50**, 7, 1039–1046, (1987).

[20] Ninfa, A. J. and Atkinson, M., R., *Trends in Microbiology*, **8**, 4, 172–179, (2000).

[21] Pedraza, J. M. and van Oudenaarden, A., *Science*, **307**, 5717, 1965–1969, (2005).

[22] Raser, J. M. and O'Shea, E. K., *Science*, **309**, 5743, 2010–2013, (2005).

[23] Reitzer, L. J. and Magasanik, B., *PNAS*, **82**, 7, (1985).

[24] Rosenfeld, N. et al., *Science*, **307**, 5717, 1962–1965, (2005).

[25] Sbano, L. and Kirkilionis, M., Analysis of a singe gene switch, *preprint*, (2006).

[26] Shiau, S. P., Schneider, B. L., Gu, W. and Reitzer, L. J., *Journal of Bacteriology*, **174**, 1, 179–185, (1992).

[27] Su, W.,Porter, S., Kustu, S. and Echols, H., *Proceedings of the National Academy of Sciences of the United States of America*, **87**, 14, 5504–5508, (1990).

[28] Suel, G. M., Garcia-Ojalvo, J., Liberman, L. M. and Elowitz, M. B., *Nature*, **440**, 7083, 545–550, (2006).

[29] Tegner, J. et al., *PNAS*, **100**, 10, 5944–5949, (2003).

[30] van Kampen, N. G., Stochastic Processes in Physics and Chemistry, (1992).

Bistability in Chemical Reaction Networks

Mirela Domijan
Mathematics Institute, University of Warwick
Coventry CV4 7 AL, United Kingdom

Abstract

Bistability, the phenomenon where two stable steady states can coexist simultaneously, is an important feature of chemical reaction networks. One step up, in the biological context, this bistability represents a type of switching mechanism, linked to most basic levels of cell operation. In very large reaction networks, it is a daunting task to identify parts of the network that could be a source of bistability. It is also not clear which components of the network might support (or destroy) this bistability once larger portions of the network are functioning. The aim of our project is to tackle these questions via stoichiometric network analysis (SNA) and graph theoretic methods. Here we will outline the latter aim of the project and present an algorithm for bistability-preserving network extensions.

Introduction

Chemical reaction networks are used to describe interactions of chemical species, where the nodes are represented by the species concentrations and links represent their interactions. The nodes are dynamic, since species concentrations can change over time. By links we think of the influence one species has on the dynamics of another. If species S_1, for example, promotes the production of S_2, then there is a link between the species. In a digraph such a link is represented by a directed arrow pointing from the species exerting the influence (S_1), to the species whose dynamics are affected (S_2).

Our focus is an investigation of dynamics (changes to links and nodes) on chemical reaction networks. More precisely, we concentrate on the phenomena of bistability. Bistability describes a situation where a network contains two stable equilibria. Recently a plethora of theoretical and numerical evidence has emerged that suggest the importance of bistability in a number of chemical reaction networks [Fe]. However, how bistability arises in a network is not yet clear [Fe] and hence modelling and a meaningful interpretation of the models are difficult. Thus, it is of interest to devise a way to identify parts of a network which create bistability, as well as to identify other parts of the network which though they themselves may not give rise to bistability, will at least preserve it.

Here we will only describe our work on the second part of this project. We will describe our bistability-preserving algorithm developed from the theory of stoichiometric network analysis, a popular analysis method in biochemical and genetic modelling. Then, we will review some graph theoretic approaches to chemical reaction systems. Graph theory and stoichiometric network analysis are closely related and one of our future aims is to translate results presented here to graph theoretic results. Graphical methods can easily be automated, hence making them natural candidates for analysis of large-scale chemical reaction networks.

Stoichiometric Network Analysis

A chemical reaction system with r reactions and m reacting species is represented by reaction equations that are derived from the reaction schemes. The following is a typical reaction,

$$\alpha_{1j}S_1 + \cdots + \alpha_{mj}S_m \to^{k_j} \beta_{1j}S_1 + \cdots + \beta_{mj}S_m, \quad j = 1, ..., r. \tag{1}$$

where S_i are the reacting species and k_i is the kinetic constant, a coefficient taking into account all other effects on the reaction rate aside from reactant concentrations, for example, temperature or light conditions, or ionic strength in the reaction.

Coefficients α_{ij} and β_{ij} represent the number of S_i molecules participating in j-th reaction at reactant and product stages. The net amount of species S_i produced or consumed by the reaction is denoted by a stoichiometric coefficient, $n_{ij} = \beta_{ij} - \alpha_{ij}$. These coefficients are arranged in a *stoichiometric* matrix, denoted N. Rate at which j-th reaction takes place, is modeled under the assumption that reactions obey mass-action kinetics and hence takes the form of a monomial,

$$v_j(x, k_j) = k_j \prod_{i=1}^{m} x^{\kappa_{ij}} \tag{2}$$

where κ_{ij} is the molecularity of the species S_i in the j-th reaction. In mass-action kinetics, kinetic exponent κ_{ij} reduces to being α_{ij}. Kinetic exponents are arranged in a *kinetic* matrix, denoted κ.

Time evolution of the species concentrations is described by the following initial value problem,

$$\dot{x} = Nv(x, k)$$
$$x(0) > 0$$

where $x(0)$ are initial species concentrations.

Steady states of the network (x_0) are determined by condition,

$$Nv(x_0, k) = 0$$

hence, all stationary fluxes belong to an intersection of the kernel of N and the positive orthant of the reaction space,

$$v(x_0) \in \left\{ z \in R^r | Nz = 0, z \in R^r_{\geq 0} \right\}$$

which form a convex polyhedral cone K_v. K_v is spanned by a set of minimal generating vectors E_i's,

$$K_v = \left\{ \sum_{i=1}^{t} j_i E_i : j_i > 0 \; \forall \, i \right\}.$$

These generating vectors or *extreme currents* decompose the network into minimal steady-state generating subnetworks. Influence of a subnetwork on full network dynamics depend on constants j_i's, called convex parameters.

The advantage of SNA is that the network stability can be inferred from certain properties of the extreme currents [Cl]. This eliminates frequently complicated computation of steady states and also does not require any knowledge of the rate constants. For example, [Cl] showed that if a network is composed entirely of extreme currents that are mixing stable (a special property referring to jacobians of subnetworks spanned by the extreme current), then this network has an asymptotically stable interior steady state for any choice of rate constants.

Graph Theory and relation to SNA

Extensive literature exists on interpretation of chemical reaction schemes via graph theory [Te]. A close relation between stoichiometric network analysis and graph theory was explored in [Gt]. Every chemical reaction network of form (1) can be represented by two graphs: a weighted directed graph that describes reactions and a bipartite undirected graph that describes reactants. For a weighted directed graph, left and right hand sides of each reaction are interpreted as complexes C_j, $j = 1, ..., n$. they form vertices of the graph. Edges of the graph (C_i, C_j) are the arrows of the reaction. They are oriented, and their weights are the reaction constants of the each reaction.

The undirected bipartite graph represents the formation of the complexes taking part in the reaction scheme. The graph has two sets of vertices, namely, a set of the complexes C_j and of the species S_i, that are linked if the complex C_j contains the species S_i. If complex C_j containing species S_i takes part in the l-the reaction, the weight on link between C_j and S_i is the coefficient α_{il}.

Each graph is associated with incidence matrices. The directed graphs has two such matrices: I_α and I_k. I_α contains the information whether the complex is the initial (-1) or end (1) vertex of an edge. I_k consists of non-zero entries only for the initial vertices, and these are the weights of the corresponding edge, i.e., the reaction constants.

The bipartite graph consists of one matrix, denoted Y. Each entry y_{ij} is the weight of the edge between the set vertices of species S_i and complexes C_j. The monomials from the reaction rates in (2) can also be obtained from the columns y_j of Y. This can be interpreted as a mapping, $\Psi_j(x) = x^{y_j}$, $j = 1, ..., n$. Then, the dynamics of the chemical

33

reaction system can be rewritten as

$$\dot{x} \;=\; Y I_\alpha I_k \Psi(x)$$

where $N = Y I_\alpha$ and $v(x,k) = I_k \Psi(x)$.

The extreme currents belong to the kernel of the product of matrices Y and I_α. [Gt2] separate extreme currents into two sets, one belonging to $Ker(I_\alpha)$, called positive circuits, and the second set belonging to $Ker(YI_\alpha)\backslash Ker(I_\alpha)$, called stoichiometric generators. Positive circuits and stoichiometric generators are closely linked to stability of network's steady states.

Extension Algorithm

Now we proceed to describe the extension algorithm for bistable networks.

Let X be a bistable network composed of n extreme currents. Following extensions preserve bistability:

1. **extensions involving species of** X. Any small extreme currents between existing species of the network are allowed.

2. **extensions involving new species not linked to** X. If an uncoupled network of new species is added to X it needs to satisfy either of the two following properties:

   (a) it is made up solely of mixing stable extremal currents, or

   (b) satisfies modified Routh-Hurwitz condition.

3. **extensions involving new species linked to** X. If interactions amongst new species are of type described in (2), then any small extreme currents between original network and network of new metabolites are allowed.

The proofs for these extensions are straight-forward. Result (1) comes from perturbation theory and is a consequence of the Implicit Function Theorem. For details on perturbation theory refer to [Gu].

To show (2), let Y be a new network of m currents not coupled to X. In order for X and Y to preserve bistability, Y must have a stable steady state. Thus we require conditions on Y. Condition (2a) comes from a result of [Cl], stated earlier. Since it requires a very strong property on Y, we also give an alternative condition (2b).

In order to sketch (2b), we need to introduce some definitions. We say that a matrix is stable if all of its eigenvalues have negative real part. Jacobian of a network evaluated at a stable steady state is a stable matrix. One way of checking matrix stability is to use the Routh-Hurwitz condition [Ga]. In SNA, [Cl] showed that all jacobians of the network belong to a family of matrices $Jac(j,h)$ that are parametrized by convex parameters j and any vector h of same dimensions as concentration space, x. Most SNA stability results come from this family of matrices $Jac(j,h)$ which is easy to compute and analyse. In fact, all jacobians of the network correspond to the $Jac(j,h)$ matrices where j and h satisfy particular conditions defined in [Gt1]. Imposing the Routh Hurwitz condition on $Jac(j,h)$

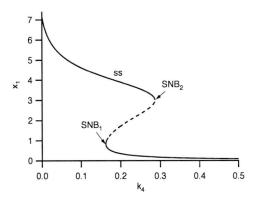

Figure 1: *Bifurcation diagram of the model for $k_1 = 0.3, k_2 = 1, k_3 = 2, k_5 = 5, k_6 = 2$. k_4 is a bifurcation parameter. Model is bistable for $k_4 \in (SNB_1, SNB_2)$. ss, curve of steady-states; SNB, saddle-node bifurcation; solid line denotes stability, dashed line denotes instability.*

together with conditions on the pair (j, h) from [Gt1] will guarantee that Y has a stable steady state.

Result (3) is a combination of (1) and (2). Given the new species form their own network Y of type described in (2), then union of X and Y is still a bistable network. Addition of small extremal currents is a result from IFT, as in (1).

Example

We now apply our algorithm. We start with a large network taking the form,

$$
\begin{aligned}
\dot{x}_1 &= -k_1 x_1^2 + k_2 + k_3 x_1 - k_4 x_1 x_2 - k_7 x_1 x_2^2 \\
\dot{x}_2 &= k_5 - k_6 x_2 - k_4 x_1 x_2 + k_7 x_1 x_2^2 - k_{13} x_2 x_4 \\
\dot{x}_3 &= k_8 - k_9 x_3 - k_{12} x_3^2 x_4 \\
\dot{x}_4 &= k_{10} x_5 - k_{11} x_4^2 - k_{12} x_3^2 x_4 - k_{13} x_2 x_4 \\
\dot{x}_5 &= k_9 x_3 - k_{10} x_5.
\end{aligned}
$$

The directed and bipartite graphs of the full network are shown in Figures 2.

Analysing the dynamics of this network is already complicated because we are dealing with a large number of parameters. Setting reaction constants k_7 to k_{13} to zero, this network reduces to a smaller subnetwork,

$$
\begin{aligned}
\dot{x}_1 &= -k_1 x_1^2 + k_2 + k_3 x_1 - k_4 x_1 x_2 \\
\dot{x}_2 &= k_5 - k_6 x_2 - k_4 x_1 x_2.
\end{aligned}
$$

35

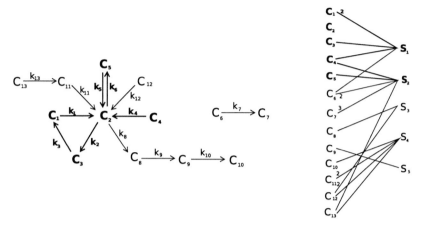

Figure 2: *Directed graph and bipartite graph of the network.*

This subnetwork displays bistability for a range of values of k parameters (Figure). It is minimal, so that further elimination of any of the terms (setting $k_i = 0$ for at least one $i = 1, .., 4$) prevents the system from displaying bistability (for details see [Se]). This subnetwork is shown by thicker lines in the directed and bipartite graphs (Figure 2).

The reduced stoichiometric matrix N_b and reaction rate vector $v_b(x_1, x_2; k_1, ..., k_4)$ of this subnetwork can be computed so to verify that this subnetwork is decomposed into further five subnetworks (Figure) spanned by extreme currents,

$$
\begin{aligned}
E_1 &= (1, 1, 0, 0, 0, 0) \\
E_2 &= (1, 0, 1, 0, 0, 0) \\
E_3 &= (0, 1, 0, 1, 1, 0) \\
E_4 &= (0, 0, 0, 0, 1, 1) \\
E_5 &= (0, 1, 1, 1, 0, 0)
\end{aligned}
$$

where $v_b(x; k) = (k_1 x_1^2, k_2, k_3 x_1, k_4 x_1 x_2, k_5, k_6 x_2)$.

Now, our aim is to determine which parts of the full network can be "switched on" (i.e., some parameters k_i for $i = 7, ..., 13$ can be increased from zero), so that the extended subnetwork, and ultimately the full network, can still preserve bistability.

Our first extension is to switch on the reaction involving k_7. We can derive the extreme currents that will span only the reaction space which is a cartesian product of the subspace $v_b(x, k)$ and coordinate v_7. We also keep in mind that every one of the extreme currents of a subnetwork is also an extreme current of an extension of this subnetwork [Cl]. This means that E_1 to E_5 are also extreme currents of the extended subnetwork with $k_7 > 0$ and all other (new) currents of this extended subnetwork will have a v_7 coordinate. Moreover, when later larger parts of the network are switched on, there might be other extreme currents that also have coordinates in the reaction rate space, $v_b(x, k) \times v_7$. By extension

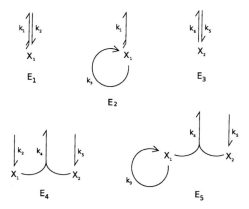

Figure 3: *Clarke's network diagram of the five extreme currents of a minimal bistable subnetwork.*

(1), given all these new currents are sufficiently small, bistability will be preserved. This implies that there exists a value of $k_7 > 0$ for which the extended network is still bistable.

Next, we look at terms k_8 to k_{12}. They take part in a subnetwork of species x_3, x_4 and x_5 that is uncoupled from species x_1 and x_2,

$$\dot{x}_3 = k_8 - k_9 x_3 - k_{12} x_3^2 x_4$$
$$\dot{x}_4 = k_{10} x_5 - k_{11} x_4^2 - k_{12} x_3^2 x_4$$
$$\dot{x}_5 = k_9 x_3 - k_{10} x_5.$$

When stoichiometric analysis is preformed on this subnetwork, it can be shown that this subnetwork consists of two mixing stable currents. From Extension (2), it follows that for any combination of k_8 to k_{12} parameters, this subnetwork (x_3, x_4, x_5) will have an asymptotically stable fixed point. Hence the extended network (with only $k_{13} = 0$) will still display bistability.

Now taking $k_{13} > 0$ couples the two subnetworks of $\{x_1, x_2\}$ and $\{x_3, x_4, x_5\}$. With reaction rate space enlargened to include coordinate v_{13}, we could calculate all the extreme currents again. New currents will have a non-zero v_{13} coordinate. Via Extension (3), when some of them are chosen to be sufficiently small (and others we could set to zero), network will still be bistable. In this case, it is possible to find $k_{13} > 0$ such that ultimately the full network is bistable.

Conclusion

Here we have presented a method for extending a bistable subnetwork such that bistability is preserved. Our algorithm does not rely on calculating the reaction constants nor on calculation of steady states. Thus it provides a fast and efficient way of extending a bistable network. Our aim is to also give a graph theoretic interpretation to our results, in the form explored by [Gt].

Bibliography

[Cl] B. Clarke, Stability of complex reaction networks, in I. Prigogine and S. Rice (Eds.), Advances in Chemical Physics, New York Wiley Vol. 43, pp 1–216, 1980.

[Fe] G. Craciun, Y. Tang, and M. Feinberg, Understanding bistability in complex enzyme-driven reaction networks, PNAS 30, Vol. 103, pp 8697–8702, 2006.

[Ga] F.R. Gantmacher, Application of the Theory of Matrices, Interscience Publishers Ltd, London, 1959.

[Gt] K. Gatermann. Counting stable solutions of sparse polynomial systems in chemistry. In: Symbolic Computation:Solving Equations in Algebra, Geometry, and Engineering. In: Contemporary Mathematics 286. American Mathematical Society, 53–69, 2001.

[Gt1] K. Gatermann, M. Eiswirth and A. Sensse, Toric ideals and graph theory to analyze Hopf bifurcations in mass-action systems, J. Symb. Comp. 40, pp 1361–1382, 2005.

[Gt2] K. Gatermann and M. Wolfrum, Bernstein's second theorem and Viro's method for sparse polynomial systems in chemistry. Adv. Appl. Math. 34(2), 252–294, 2005.

[Gu] J. Guckenheimer and P. Holmes, Nonlinear Oscillations, Dynamical Systems and Bifurcations of Vector Fields, Springer Verlag: Applied Mathematics Sciences 42, 2002.

[Te] O.N. Temkin, A.V. Zeigarnik and D.G. Bonchev, Chemical Reaction Networks: A Graph-Theoretical Approach, CRC Press, 1996.

[Se] A. Sensse, Convex and toric geometry to analyze complex dynamics in chemical reaction systems, PhD thesis, Otto-von-Guericke-Universitat Magdeburg, 2005.

Genetic Networks

From DNA arrays to modules to models

Sven Bergmann
Department of Medical Genetics, University of Lausanne
CH-1015 Lausanne, Switzerland

Abstract

DNA microarrays have become one of the most prominent data sources in the fields of bioinformatics and cellular systems biology. This high-throughput technology allows measuring transcript levels simultaneously for thousands of genes. Thus collections of many microarrays elucidate gene regulation under many experimental conditions and can provide insight into the entire transcription network. Here we review the methodolgy for identifying *transcription modules* of co-expressed genes and the experimental conditions in which they are co-expressed, from large datasets of expression data. These provide the basic building blocks of the genome-wide transcription program, as well as a map characterizing the transcriptional changes induced by novel experiments.

Introduction

DNA microarrays have firmly established themselves as a standard tool in biological and biomedical research. Together with the rapid advancement of genome sequencing projects, microarrays and related high-throughput technologies have been key factors in the study of the more global aspects of cellular systems biology [1]. While genomic sequence provides an inventory of parts, a proper organization and eventual understanding of these parts and their functions requires comprehensive views also of the regulatory relations between them [2]. Genome-wide expression data offer such a global view by providing a simultaneous read-out of the mRNA levels of all (or many) genes of the genome.

Most microarray experiments are conducted to address specific biological issues (Figure a). In the simplest case, such a study may focus on the expression response to the deletion of individual genes or to specific cellular conditions. Already when extending the experimental setup to include several conditions, e.g. time points along the cell-cycle [3] or several tissue samples, the sheer amount of data points necessitates computational tools

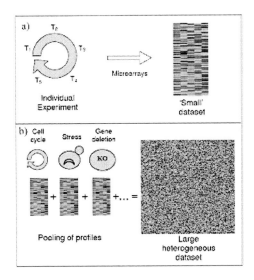

Figure 1: *Expression data* **a** Individual microarray experiments addressing specific biological issues give rise to small datasets comprising only few distinct experimental conditions (e.g. time points). **b** Large-scale expression data can be generated by pooling profiles from many such individual experiments (or conducting dedicated comprehensive assays). Such data cover not only thousands of genes but also many cellular states by including a heterogeneous collection of experimental conditions.

to extract and organize relevant biological information. A wide range of approaches have been developed, including numerous clustering algorithms, statistical methods for detecting differential expression, and dimension-reduction techniques (reviewed by Brazma and Vilo [4] and Slonim [5]).

In addition to the specific biological questions probed in such individual focused experiments, it is widely recognized that a wealth of additional information can be retrieved from a large and heterogeneous dataset describing the transcriptional response to a variety of different conditions [2]. Also the relatively high level of noise in these data can be dealt with most effectively by combining many arrays probing similar conditions. Comprehensive data have been used to provide functional links for unclassified genes [3, 6, 7, 8, 9], to predict novel cis-regulatory elements [7, 10, 11, 12] and to elucidate the structure of the transcriptional program [12, 13].

Large-scale expression data may result from systematic efforts to characterize a range of transcription states by testing many different biological conditions [6, 13, 14]. In addition, large datasets can be assembled by collecting expression profiles and pooling them into one comprehensive database (Figure b). Until recently, these data appeared in different formats and were scattered among various internet sites (if available at all) [15].

The increasing availability of microarray technology and the ensuing explosion of available expression profiles (usually obtained in different laboratories using different array technologies) have prompted the establishment of standardized annotations such as the MIAME [16] and MAGE-ML [17] standards, and a number of public repositories for chip data [18, 19, 20, 21].

Single microarray experiments are global only in the sense that the genes probed span all or most of the genome. The idea of composing large-scale expression datasets is to include a large variety of conditions in order to span also the space of transcriptional states of the cell. While this is a necessary step towards the elucidation of the transcription programs, such data present new and serious challenges to the mathematical and computational tools used to analyze them. In particular, the context-specific nature of regulatory relationships poses a difficult computational problem. Consequently, a sizeable variety of different approaches have been proposed in the literature (see review by Ihmels and Bergmann [22]).

The modular concept

Whenever we face a large number of individual elements that have heterogeneous properties, grouping elements with similar properties together can help to obtain a better understanding of the entire ensemble. For example, we may attribute human individuals of a large cohort to different groups based on their sex, age, profession, etc., in order to obtain an overview over the cohort and its structure. Similarly, individual genes can be categorized according to their properties to obtain a global picture of their organization in the genome. Evidently, in both cases alike, the assignment of the elements to groups - or modules - depends on which of their properties are considered and on how these properties are processed in order to associate different elements with the same module. A major advantage of studying properties of modules, rather than individual elements, relies on a basic principle of statistics: The variance of an average decreases with the number N of (statistical) variables used to compute its value like 1/N, because fluctuations in these variables tend to cancel each other out. Thus mean values over the elements of a module or between the elements of different modules are more robust measures than the measurements of each single element alone. This is particularly relevant for the noisy data produced by chip-based high-throughput technologies.

Co-classification of genes and conditions

The central problem in the analysis of large and diverse collections of expression profiles lies in the context-dependent nature of co-regulation. Usually genes are coordinately regulated only in specific experimental contexts, corresponding to a subset of the conditions in the dataset. Such conditions could be different cellular environments (external conditions), as well as distinct tissues or developmental stages (internal conditions).

To take these considerations into account, expression patterns must be analyzed with respect to specific subsets; genes and conditions should be co-classified [7, 28, 29, 30, 26,

27, 28]. The resulting 'transcription modules' (another common term is 'bicluster') consist of sets of co-expressed genes together with the conditions over which this co-expression is observed. The naive approach of evaluating expression coherence of all possible subsets of genes over all possible subsets of conditions is computationally infeasible, and most analysis methods for large datasets seek to limit the search space in an appropriate way. For example, Getz et al. [30] introduced a variant of biclustering based on the idea to perform standard clustering iteratively on genes and conditions. Their Coupled-Two-Way-Clustering procedure is initialized by separately clustering the genes and conditions of the full matrix. Each combination of the resulting gene and condition clusters defines a submatrix of the expression data. Instead of considering all possible combinations, two-way clustering is then applied to all such submatrices in the following iteration. Other biclustering methods, like the Plaid Model [26] and Gene Shaving [27], aim to identify only the most dominant bicluster in the dataset, which is then masked in a subsequent run to allow for the identification of new clusters. The SAMBA (Statistical-Algorithmic Method for Bicluster Analysis) biclustering method [28] combines graph theory with statistical data modeling. While each method has its advantages and disadvantages [22, 29], a common drawback is their scaling properties in terms of CPU time and memory usage when applied to large data. We developed the Signature Algorithm [7] (Figure) and an iterative extension [29, 30] of it for modular analysis of large-scale expression data. Our methods have been shown to compete well with others in terms of efficiency [31]. Moreover, because we do not compute correlations, the computation time of our algorithm scales extremely well with the size of the data (essentially linear).

From modules to models

Transcription modules provide not only the basic building blocks that characterize the structure of the genome-wide transcription program under a variety of conditions, they also supply a map for a more interpretable characterization of transcriptional changes induced by novel experiments. In particular, searching for coherent changes in expression in larger modules, one may identify patterns that are too weak to discern when considering each of its genes alone. For example, Mootha et al. [32] showed that the coordinate expression of a set of functionally related genes was significantly altered in human diabetic muscle, even though this effect was too subtle to be apparent at the single gene level (see, however comment in Ref. [33]). Segal et al. [34] used expression data from almost 2,000 published microarrays from human tumors to establish a compendium of modules combining genes with similar behavior across arrays. This cancer module map allowed them to characterize clinical conditions (like tumor stage and type) in terms of a profile of activated and deactivated modules. For example, they found that a Growth Inhibitory Module, consisting primarily of growth suppressors, was coordinately suppressed in a subset of acute leukemia arrays, suggesting a possible mechanism for the uncontrolled proliferation in these tumors.

These and other results [36, 36, 37, 38] illustrate the value of analyzing the complex processes underlying biological conditions such as human diseases in terms of transcrip-

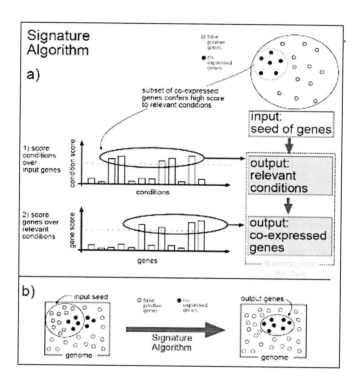

Figure 2: The Signature Algorithm requires as input a set of genes, some of which are expected to be co-regulated based on additional biological information such as a common promoter binding motifs or functional annotation. **a** The algorithm proceeds in two steps: In the first step, this input seed is used to identify the conditions that induce the highest average expression change in the input genes. Only conditions with a score above some threshold are selected. In the second stage of the algorithm, genes that are highly and consistently expressed over these conditions are identified. The result consists of a set of co-regulated genes together with the regulating conditions and is termed transcription module. **b** The output contains only the co-regulated part of the input seed, as well as other genes that were not part of the original input but display a similar expression profile over the relevant conditions.

tion modules. Yet, while a modular characterization of such processes provides a powerful tool to elucidate aspects of the normal and defective regulatory mechanisms, it is only one step towards the goal of obtaining detailed mechanistic models of the processes pertaining to disease. Since cellular processes are regulated at all stages leading from DNA to functional proteins, the integration of information on regulatory sequence, as well as post-transcriptional regulation is crucial in this endeavor.

Bibliography

[1] Kitano, H. Systems biology: a brief overview. Science 295, 1662-4 (2002).

[2] Lander, E. S. Array of hope. Nat Genet 21, 3-4 (1999).

[3] Tavazoie, S., Hughes, J. D., Campbell, M. J., Cho, R. J. , Church, G. M. Systematic determination of genetic network architecture. Nat Genet 22, 281-5 (1999).

[4] Brazma, A. , Vilo, J. Gene expression data analysis. FEBS Lett 480, 17-24 (2000).

[5] Slonim, D. K. From patterns to pathways: gene expression data analysis comes of age. Nat Genet 32 Suppl, 502-8 (2002).

[6] Hughes, T. R. et al. Functional discovery via a compendium of expression profiles. Cell 102, 109-26 (2000).

[7] Ihmels, J. et al. Revealing modular organization in the yeast transcriptional network. Nat Genet 31, 370-7 (2002).

[8] Wu, L. F. et al. Large-scale prediction of Saccharomyces cerevisiae gene function using overlapping transcriptional clusters. Nat Genet 31, 255-65 (2002).

[9] Kim, S. K. et al. A gene expression map for Caenorhabditis elegans. Science 293, 2087-92 (2001).

[10] Bussemaker, H. J., Li, H. , Siggia, E. D. Regulatory element detection using correlation with expression. Nat Genet 27, 167-71 (2001).

[11] Hughes, J. D., Estep, P. W., Tavazoie, S. , Church, G. M. Computational identification of cis-regulatory elements associated with groups of functionally related genes in Saccharomyces cerevisiae. J Mol Biol 296, 1205-14 (2000).

[12] Wang, W., Cherry, J. M., Botstein, D. , Li, H. A systematic approach to reconstructing transcription networks in Saccharomycescerevisiae. Proc Natl Acad Sci U S A 99, 16893-8 (2002).

[13] Gasch, A. P. et al. Genomic expression programs in the response of yeast cells to environmental changes. Mol Biol Cell 11, 4241-57 (2000).

[14] Su, A. I. et al. A gene atlas of the mouse and human protein-encoding transcriptomes. Proc Natl Acad Sci U S A 101, 6062-7 (2004).

[15] Brazma, A., Robinson, A., Cameron, G. , Ashburner, M. One-stop shop for microarray data. Nature 403, 699-700 (2000).

[16] Brazma, A. et al. Minimum information about a microarray experiment (MIAME)-toward standards for microarray data. Nat Genet 29, 365-71 (2001).

[17] Spellman, P. T. et al. Design and implementation of microarray gene expression markup language (MAGE-ML). Genome Biol 3, RESEARCH0046 (2002).

[18] Ball, C. A. et al. The Stanford Microarray Database accommodates additional microarray platforms and data formats. Nucleic Acids Res 33, D580-2 (2005).

[19] Brazma, A. et al. ArrayExpress–a public repository for microarray gene expression data at the EBI. Nucleic Acids Res 31, 68-71 (2003).

[20] Edgar, R., Domrachev, M. , Lash, A. E. Gene Expression Omnibus: NCBI gene expression and hybridization array data repository. Nucleic Acids Res 30, 207-10 (2002).

[21] Ikeo, K., Ishi-i, J., Tamura, T., Gojobori, T. , Tateno, Y. CIBEX: center for information biology gene expression database. C R Biol 326, 1079-82 (2003).

[22] Ihmels, J. H. , Bergmann, S. Challenges and prospects in the analysis of large-scale gene expression data. Brief Bioinform 5, 313-27 (2004).

[23] Bittner, M., Meltzer, P. , Trent, J. Data analysis and integration: of steps and arrows. Nat Genet 22, 213-5 (1999).

[24] Cheng, Y. , Church, G. M. Biclustering of expression data. Proc Int Conf Intell Syst Mol Biol 8, 93-103 (2000).

[25] Getz, G., Levine, E. , Domany, E. Coupled two-way clustering analysis of gene microarray data. Proc Natl Acad Sci U S A 97, 12079-84 (2000).

[26] Lazzeroni, L. , Owen, A. Plaid models for gene expression data. Technical report, Stanford University, Statistics (1999).

[27] Hastie, T. et al. 'Gene shaving' as a method for identifying distinct sets of genes with similar expression patterns. Genome Biol 1, RESEARCH0003 (2000).

[28] Tanay, A., Sharan, R. , Shamir, R. Discovering statistically significant biclusters in gene expression data. Bioinformatics 18 Suppl 1, S136-44 (2002).

[29] Ihmels, J., Bergmann, S. , Barkai, N. Defining transcription modules using large-scale gene expression data. Bioinformatics 20, 1993-2003 (2004).

[30] Bergmann, S., Ihmels, J. , Barkai, N. Iterative signature algorithm for the analysis of large-scale gene expression data. Phys Rev E Stat Nonlin Soft Matter Phys 67, 031902 (2003).

[31] Prelic, A. et al. A systematic comparison and evaluation of biclustering methods for gene expression data. Bioinformatics 22, 1122-9 (2006).

[32] Mootha, V. K. et al. PGC-1alpha-responsive genes involved in oxidative phosphorylation are coordinately downregulated in human diabetes. Nat Genet 34, 267-73 (2003).

[33] Damian, D. , Gorfine, M. Statistical concerns about the GSEA procedure. Nat Genet 36, 663; author reply 663 (2004).

[34] Segal, E., Friedman, N., Koller, D. , Regev, A. A module map showing conditional activity of expression modules in cancer. Nat Genet 36, 1090-8 (2004).

[35] Lamb, J. et al. A mechanism of cyclin D1 action encoded in the patterns of gene expression in human cancer. Cell 114, 323-34 (2003).

[36] Rhodes, D. R. et al. Large-scale meta-analysis of cancer microarray data identifies common transcriptional profiles of neoplastic transformation and progression. Proc Natl Acad Sci U S A 101, 9309-14 (2004).

[37] Chang, C. F., Wai, K. M. , Patterton, H. G. Calculating the statistical significance of physical clusters of co-regulated genes in the genome: the role of chromatin in domain-wide gene regulation. Nucleic Acids Res 32, 1798-807 (2004).

[38] Desai, K. V. et al. Initiating oncogenic event determines gene-expression patterns of human breast cancer models. Proc Natl Acad Sci U S A 99, 6967-72 (2002).

Gene Regulatory Networks

Yonatan Bilu
Weizmann Institute of Science
Rehovot 76100, Israel

Abstract
Gene regulatory networks are among the most extensively studied subjects in Systems Biology. This report gives a brief overview of the work done in the field, focusing on three key concepts - modules in gene regulatory networks, elements of regulation and comparative analysis of gene regulatory networks. We give only a bird's eye view of the extensive research on these issues, but aim to describe the available data sources, some of the relevant algorithms, and some specific results, as well.

Introduction

For over a decade the field of molecular biology has witnessed a shift in focus from the identification and study of individual genes and proteins to the ambitious task of understanding genetic networks as a whole, and the interactions between their components. This has been possible due to the accumulation of much data at the single-component level, and, most importantly, via technological advancements that allow for high throughput analysis of biological data, such as genome sequencing and gene expression measurements.

Indeed, both gene sequence data and gene expression data accumulated over the past years in huge numbers, and is expected to continue growing at a rapid pace. For example, currently the NCBI GenBank database [2] holds over 54 million gene sequences, Complete genomes maps are available for nearly 700 different organisms [3], and microarray databases such as SMD [4] contain expression data from tens of thousands of experiments. Nonetheless, while this data, which constitutes the elementary building blocks of genetic networks, is abundantly available, deciphering the function of genes and understanding the connections and relations among them - the design principles and topology of the genetic network - remains a challenging task.

Here we focus on the regulatory aspect of genetic networks, that is, on the regulatory relations among genes. We present the difficulties and pitfalls in unveiling these aspects of genetic networks, and some of the techniques developed for doing so. For a more complete overview of gene regulatory networks, see for example [1].

Identifying clusters in genetic networks

The basic paradigm in computational molecular biology is that of sequence homology - genes with similar sequence have similar function. Consequentially, tools for computing sequence homology were developed [5, 6, 7, 8], and databases of homology-based gene clusters were constructed [9, 10, 11, 12]. This view of the genetic network is useful for inferring gene function from one organism to another, or among a family of genes within an organism, but is clearly limited in its ability to expand our knowledge beyond its contemporary state.

DNA microarray technology allows us to go beyond gene sequence - into the phenotypic behavior of genes (cf. [13]). It allows measuring transcript levels for many genes - in many cases, for *all* genes in the genome - shedding light on their expression patterns in different tissues, or as a response to different conditions. Hence, it is a key tool in unveiling the topology of gene regulatory networks [14, 15, 16]. In particular, as with sequence homology, it is plausible to deduce that genes with similar expression patterns are related in their function. Moreover, an understanding of this function may be gained by studying these patterns - genes whose expression levels vary significantly from their base level under certain conditions are likely to be involved in the cell response to these condition. Similarly, genes with distinct expression patterns in a specific tissue are likely to be involved with the function of that tissue.

Genetic networks are usually constructed from expression data by co-expression. In the terminology of the introductory chapter this is a static, undirected network with weights attached to the links: The nodes correspond to genes, and the weight attached to the link (g_1, g_2), is the correlation coefficient for the expression of both genes over the set of experiments. In other words, the weight of the link represents the likelihood that the genes are co-regulated and involved in similar functions. A more refined representation allows looking at temporal changes in the weights, either during time series or varying experimental conditions.

Clustering genes according to their expression patterns has been the subject of much work (see [22] for a comprehensive survey). Roughly, this corresponds to finding cliques of nodes with "heavy" links among them in the network described above. Earlier works focused on specific sets of experimental conditions, and identifying genes related to them [17, 18, 19, 20, 21]. But as the amount of available expression data grew, it became important to identify not only sets of co-expressed genes, but also sets of related conditions, under which these genes are likely to function.

One such approach is the signature algorithm [23], which uses a bi-clustering technique to identify "transcription modules" - sets of co-regulated genes alongside sets of conditions in which they are co-regulated. Importantly, these modules may overlap, reflecting the well-known flexibility of genetic networks, where a specific gene is often involved in several, possibly unrelated, functions. Further improvements on this algorithm - the iterative signature algorithm [24] and the progressive iterative signature algorithm [25] - delve into

49

the topology of the genetic network by constructing a hierarchy of transcription modules, based on refined clustering criteria.

Another promising approach is that of modeling the data as a Bayesian network (reviewed in [26]). One thinks of the expression level of a gene g in experiment e as the value of a random variable $X_{g,e}$. Given a partition of the gene set and the experimental conditions set, one can model the dependencies among the variables in a concise way, via a Bayesian Network, and infer their conditional probabilities. Moreover, using an Expectation-Maximization algorithm, one can learn the partition of the genes and experimental conditions that best fits the data. The advantage of this method is that it describes the topology of the network in a "soft", probabilistic way, in a well-defined statsitical setting, where as a hierarchical topology is deterministic, and cutoff conditions are somewhat arbitrary. By contrast, this method partitions the genes and experimental conditions into disjoint clusters, while the Signature Algorithm has the advantage of producing a "soft" clustering.

An important issue in assessing the numerous gene expression clustering algorithms is a benchmark for validating their predictions. The *de facto* tool for this in recent years is the Gene Ontology (GO, [27]) database, which assigns functional attributes - biological process, molecular function and cellular component - to many of the known genes. Clusterings are evaluated according to the homogeneity of the appropriate GO terms assigned to their constituent genes. Indeed, the Iterative Signature Algorithm, for example, when applied to the yeast *S. Cerevisiae*, identifies at low resolution five modules, corresponding to its central functions - protein synthesis, stress response, cell cycle, mating response and amino acid biosynthesis [24].

Elements of Regulation

Regulation of gene products, especially in eukaryotes, is done at multiple levels: the accessibility of genes to the transcriptional machinery, the lifetime of the mRNA product and its accessibility to the translational machinery, and also the folding and trafficking of the resulting protein [28]. Of these, the most prominent method for regulation is at the transcriptional level, via proteins known as transcription factors (TFs).

Understanding the topology of regulatory networks relies therefore, to a large extent, on identifying genes which act as TFs, and, more importantly, on identifying the targets of these TFs. Roughly put, the genomic data defines the nodes of the network in a manner which is relatively easy to comprehend. The difficulty is in identifying the edges of the network - the regulatory relations among genes.

Transcription factors regulate gene expression by binding to the so-called "promoter region" of a gene - a region of the DNA adjacent to that of the gene - and manipulating the action of the RNA polymerase machinery. For example, this may be done by recruiting the protein components of this machinery (an "activator"), or by blocking them (an "inhibitor"). Transcription factors bind to sequence-specific sites within the promoter region, known as Transcription Factor Binding Sites (TFBSs) or *cis*-regulatory elements (CREs). In higher Eukaryotes, these elements often exhibit a higher-order structure known as *cis*-regulatory modules (CRMs) [28].

While identifying genes along a genomic sequence is by now fairly routine, and can be done computationally with high accuracy, identifying CREs is a difficult task. Unlike genes, CREs are very short, often no more than 10bp long. Moreover, they are often "fuzzy", that is, different CREs bound by the same TF have similar - but usually not identical - sequences, and again unlike genes, may incorporate "out-of-frame" modifications without loosing functionality. Hence, it is expected that chance occurrences of genomic sub-sequences similar to those recognized by a TF will not be rare, and, indeed, many of them are spurious.

This complexity is also reflected in the state of available data. The most comprehensive database for TFs and their targets is the TRANSFAC [29] commercial database. Aside from the fact that its free-of-charge version is less maintained and data reduced, even the full version is limited in its ability to reconcile the variability that exists in suggested TF binding sequences.

The main computational approach in identifying CREs is by scanning the promoter regions of genes which are thought to be regulated by the same transcription factor, and looking for subsequences for which these regions are enriched (see [30, 31] for a review). This set of genes is usually taken to be a cluster of co-expressed genes, as discussed in the previous section. In fact, one of the criteria for evaluating the quality of gene expression clustering algorithms is in their ability to recognize known CREs this way (e.g. in [24]). A more sophisticated implementation of this approach can be used to identify the co-dependencies inherent to transcriptional regulation [32]. Notably, these approaches require a definition of the genes' promoter region, which, in higher eukaryotes, is often not well understood.

One way to complement this basic approach is by considering sequence conservation - it is plausible to assume that CREs will tend to be conserved among closely related species. Hence, subsequences for which promoter regions are enriched in several related species, especially if they are located at approximately the same position within these regions, are particularly likely to represent true CREs. See [33, 34] for a review of such methods.

Co-expression of genes, while indicative of regulation by the same TFs, is certainly not clear evidence of this, especially when combinatorial regulation and effects upstream in the network are considered. A more direct identification of the promoter regions bound by a given transcription factor is offered by ChIP-CHIP essays [35]. In these, chromatin immuno-precipitation (ChIP) is used to seclude all DNA segments bound by a specific TF, and microarrays are used to identify these segments. A combination of this methodology with the computational approach above has led to a the construction of the first comprehensive map of CREs in a eukaryote (albeit yeast) [36].

Notably, this technology, when combined with gene expression data, is also useful in constructing clusters of co-regulated gene. The GRAM algorithm [37] identifies clusters of genes which are not only co-expressed, but also exhibit the same TFs bound to the promoter regions. Indeed, such modules revealed unexpected regulatory connections, such as stress response factors in yeast being involved in the pheromone response pathway.

ChIP-CHIP methodology allows for high throughput identification of targets of a given TF, but requires much work for each TF. Furthermore, gene regulation is clearly an

51

adaptive response, and hence to decipher the binding sites of a TF, measurements need to be done in all relevant conditions. Hence, it is not clear when this methodology will be developed enough to yield a comprehensive CRE map of a higher eukaryote.

Another method for identifying a TF's targets is by deleting its coding gene [38, 39], or over-expressing it, and observing the genome-wide changes in expression patterns. The downside of this approach is that it is difficult to distinguish between direct and indirect effects. Moreover, it is not useful if the TF is essential for viability, and becomes infeasible to conduct comprehensively when combinatorial effects are sought, due the huge amount of possible strains.

Recently, the important role of micro-RNAs (miRs) in the regulatory networks of higher eukaryotes has become apparent. Micro-RNAs compose a post-transcriptional regulatory mechanism, by binding to specific sequences in the 3' UTR of mRNA, which either prevents it from being translated, or even leads to its degradation. It is speculated that 1% of human genes are miRs, regulating the level of 10% of the genes. See [40] for a recent review of miRs.

Databases for miRs include miRBase [41], which lists nearly 4000 miRs (and their targets), and miRNAMap [42], which is focused on mammalian miRs, and lists over 2000 of them for Human, Mouse, Rat and Dog alone. Notably, many of these miRs and miR-targets were identified using computational methods, so their fidelity is not clear.

Comparative analysis of gene regulatory networks

The abundance of fully-sequenced genomes is not only a mean for corroborating sequence analysis, but also allows for a comparison of the genetic networks encoded in these genomes. In particular, it may reveal the evolutionary dynamics of such networks.

The basic characteristics of the topology of genetic networks of different organisms is surprisingly similar, and the expression of individual genes, as reflected by co-expression of orthologous genes, is often conserved. One example of this is the distribution of "network motifs" [35] in the network, which is found to be similar in *E. coli*, *B. subtilis* and *S. cerevisiae*. Another is the comparison of the networks constructed from gene expression data of *S. cerevisiae*, *C. elegans*, *E. coli*, *A. thaliana*, *D. melanogaster*, and *H. sapiens*. Albeit the large evolutionary distance among them, all display a connectivity distribution which follows a power-law, higher essentiality and conservation of highly connected genes, and high modularity of the expression program [44]. Yet another example is that perturbations of the network in *C. elegans* and *D. melanogaster*, which lead to similar gene-expression reactions [49].

As with sequence homology and gene co-expression, computational tools were developed to align genetic networks from different organisms. That is, a global comparison of the networks aims to identify the similar (conserved) sub-networks, and those sub-networks which have differentiated [45]. Another approach is that of "network integration", where data of different types over the same set of genes is combined to study gene or protein modules [46, 47]. Gene modules, or sub-networks, can also be sought for directly, using techniques for "network querying" [45]. See [48] for a review of these methods.

Identifying the differences among networks, and relating them to the biological differences between the organism, is, probably, the most challenging question in this field. A striking example of how this can indeed be achieved, was given by the comparative analysis of the transcriptional program of *S. cerevisiae* and *C. albicans* [50]. A *Differential Clustering Algorithm* was used to trace the composure of the transcription modules, identifying those which are conserved in both organisms, and those which are split in one of them. For example, in *C. albicans* genes involved in cytoplasmatic and mitochondrial translation are co-expressed, forming one module. In *S. cerevisiae* they form distinct modules. Indeed, rapidly growing *S. cerevisiae* cells utilize fermentation and do not require oxygen. In contrast, rapid growth in *C. albicans* relies on aerobic respiration and requires mitochondrial functions.

To understand the finer differences among genetic networks it is illuminating to study closely related species, or even strains, and map phenotypic expression differences to the underlying genomic ones. Such studies have suggested that, indeed, evolution may depend more strongly on variation in gene expression, that is, on the topology of the genetic regulatory network, than on differences between variant forms of proteins.

For example, 86 segregants of laboratory and wild strains of *S. cerevisiae*, were studied using linkage analysis to map genomic changes to co-expressed gene clusters [51]. Interestingly, Most changes were mapped to *trans*-acting loci, that is, to a location distinct from the promoter regions of the co-expressed genes. Surprisingly, there was no significant enrichment for transcription factor coding genes among these loci, stressing the complexity and multi-layered nature of expression regulation.

Acknowledgements
I would like to thank Naama Barkai for fruitful discussions.

Bibliography

[1] T. G. Dewey and D. J. Galas. Gene Regulatory Networks in Power Laws, Scale-Free Networks and Genome Biology, eds. Koonin, E., Wolf, Y., and Karev, G., Landes Bioscience/Eureka.com, Georgetown, TX (2004).

[2] Benson DA, Karsch-Mizrachi I, Lipman DJ, Ostell J, Wheeler DL. GenBank. Nucleic Acids Res. 2005 Jan 1;33.

[3] http://www.ncbi.nlm.nih.gov/entrez/query.fcgi?db=genome.

[4] Ball CA, Awad IA, Demeter J, Gollub J, Hebert JM, Hernandez-Boussard T, Jin H, Matese JC, Nitzberg M, Wymore F, Zachariah ZK, Brown PO, Sherlock G. The Stanford Microarray Database accommodates additional microarray platforms and data formats. Nucleic Acids Res 2005 Jan 1;33(1).

[5] Altschul, SF, Gish, W, Miller, W, Myers, EW, and DJ Lipman (1990). Basic local alignment search tool. J. of Mol. Biol. 215:403-10.

[6] Altschul, SF, Madden, TL, Schaffer, AA, Zhang, J, Zhang, Z, Miller, W, and DJ Lipman (1997). Gapped BLAST and PSI-BLAST: a new generation of protein database search programs. Nucleic Acids Res. 25(17):3389-402.

[7] Pearson, W.R. (1990) Rapid and Sensitive Sequence Comparison with FASTP and FASTA. Methods in Enzymology 183:63-98.

[8] Thompson, J.D., Higgins, D.G. and Gibson, T.J. (1994) CLUSTAL W: improving the sensitivity of progressive multiple sequence alignments through sequence weighting, position specific gap penalties and weight matrix choice. Nucl. Acids Res. 22:4673-4680.

[9] Gasteiger E., Jung E., Bairoch A. Swiss-Prot: connecting biological knowledge via a protein database. Curr. Issues Mol. Biol. 3:47-55(2001).

[10] Wu CH, et al. The Universal Protein Resource (UniProt): an expanding universe of protein information. Nucleic Acids Res. 2006 Jan 1;34.

[11] Yona G, Linial N, Linial M. ProtoMap: automatic classification of protein sequences and hierarchy of protein families. Nucleic Acids Res. 2000 Jan 1;28(1):49-55.

[12] Kaplan N, Sasson O, Inbar U, Friedlich M, Fromer M, Fleischer H, Portugaly E, Linial N, Linial M. ProtoNet 4.0: a hierarchical classification of one million protein sequences. Nucleic Acids Res. 2005 Jan 1;33.

[13] Lander, E. S. Array of hope, Nature Genet., Vol. 21, pp. 3-4. 1999.

[14] Tavazoie, S., Hughes, J. D., Campbell, M. J. et al. (1999), Systematic determination of genetic network architecture, Nature Genet., Vol. 22, pp. 281-285.

[15] Brazma, A. and Vilo, J. (2000), Gene expression data analysis, FEBS Lett., Vol. 480, pp. 17-24.

[16] Slonim, D. K. (2002), From patterns to pathways: Gene expression data analysis comes of age, Nature Genet., Vol. 32(Suppl.), pp. 502-508.

[17] Eisen MB, Spellman PT, Brown PO and Botstein D. (1998). Cluster Analysis and Display of Genome-Wide Expression Patterns. Proc Natl Acad Sci U S A 95, 14863-8.

[18] Chu, S. et al. The transcriptional program of sporulation in budding yeast. Science 282, 699-705 (1998).

[19] DeRisi, J.L., Iyer, V.R. and Brown, P.O. Exploring the metabolic and genetic control of gene expression on a genomic scale. Science 278, 680-686 (1997).

[20] Gasch, A. P., Spellman, P. T., Kao, C. M. et al. (2000), Genomic expression programs in the response of yeast cells to environmental changes, Mol. Biol. Cell, Vol. 11, pp. 4241-4257.

[21] U. Alon, N. Barkai, D.A. Notterman, K. Gish, S. Ybarra, D. Mack, A. J. Levine. Broad Patterns of Gene Expression Revealed by Clustering Analysis of Tumor and Normal Colon Tissues Probed by Oligonucleotide Arrays. Proc. Natl. Acad. Sci., USA , 96,6745-50 (1999)

[22] Ihmels J, Bergmann S. Challenges and prospects in the analysis of large-scale gene expression data. Brief Bioinform. 2004 Dec;5(4):313-27.

[23] Ihmels J, Friedlander G, Bergmann S, Sarig O, Ziv Y, Barkai N. Revealing modular organization in the yeast transcriptional network Nat Genet. 2002 Aug;31(4):370-7. Epub 2002 Jul 22.

[24] Bergmann S, Ihmels J, Barkai N. Iterative signature algorithm for the analysis of large-scale gene expression data. Phys Rev E Stat Nonlin Soft Matter Phys. 2003 Mar;67(3 Pt 1):031902. Epub 2003 Mar 11.

[25] Kloster M, Tang C, Wingreen NS. Finding regulatory modules through large-scale gene-expression data analysis. Bioinformatics. 2005 Apr 1;21(7):1172-9.

[26] Friedman, N. Inferring cellular networks using probabilistic graphical models. Science. 2004 Feb 6;303(5659):799-805.

[27] Gene Ontology: tool for the unification of biology. The Gene Ontology Consortium (2000) Nature Genet. 25: 25-29.

[28] Carey, M. and S.T. Smale, Transcriptional regulation in Eukaryotes. 1999, Cold Spring Harbor, New York: CSHL Press.

[29] Matys V. et al. TRANSFAC: transcriptional regulation, from patterns to profiles. Nucleic Acids Res. 2003 Jan 1;31(1):374-8.

[30] Stormo GD. DNA binding sites: representation and discovery. Bioinformatics. 2000 Jan;16(1):16-23.

[31] Ohler U, Niemann H. Identification and analysis of eukaryotic promoters: recent computational approaches. Trends Genet 2001, 17:56-60.

[32] Pilpel, Y., Sudarsanam, P. and Church, G. Identifying regulatory networks by combinatorial analysis of promoter elements. Nature Genetics 29 153-159 (2001).

[33] Lenhard B, Sandelin A, Mendoza L, Engstrom P, Jareborg N, Wasserman WW. Identification of conserved regulatory elements by comparative genome analysis. J Biol 2003, 2:13.

[34] Nardone J, Lee DU, Ansel KM, Rao A. Bioinformatics for the 'bench biologist': how to find regulatory regions in genomic DNA. Nat Immunol. 2004 Aug;5(8):768-74.

[35] Lee TI et al. Transcriptional regulatory networks in Saccharomyces cerevisiae. Science. 2002 Oct 25;298(5594):799-804.

[36] Harbison CT, Gordon DB et al. Transcriptional regulatory code of a eukaryotic genome. Nature. 2004 Sep 2;431(7004):99-104.

[37] Bar-Joseph Z, Gerber GK, Lee TI, Rinaldi NJ, Yoo JY, Robert F, Gordon DB, Fraenkel E, Jaakkola TS, Young RA, Gifford DK. Computational discovery of gene modules and regulatory networks. Nat Biotechnol. 2003 Nov;21(11):1337-42.

[38] Giaever, G. et al. Functional profiling of the Saccharomyces cerevisiae genome. Nature. 2002 Jul 25;418(6896):387-91.

[39] Bensen RJ, Johal GS, Crane VC, Tossberg JT, Schnable PS, Meeley RB, Briggs SP. Cloning and characterization of the maize An1 gene. Plant Cell. 1995 Jan;7(1):75-84.

[40] Carthew RW. Gene regulation by microRNAs. Curr Opin Genet Dev. 2006 Apr;16(2):203-8.

[41] Griffiths-Jones S, Grocock RJ, van Dongen S, Bateman A, Enright AJ. miRBase: microRNA sequences, targets and gene nomenclature. Nucleic Acids Res. 2006 Jan 1;34. 140-4.

[42] Hsu PW, Huang HD, Hsu SD, Lin LZ, Tsou AP, Tseng CP, Stadler PF, Washietl S, Hofacker IL. miRNAMap: genomic maps of microRNA genes and their target genes in mammalian genomes. Nucleic Acids Res. 2006 Jan 1;34. 135-9.

[43] Milo R, Shen-Orr S, Itzkovitz S, Kashtan N, Chklovskii D, Alon U. Network motifs: simple building blocks of complex networks. Science. 2002 Oct 25;298(5594):824-7.

[44] Bergmann S, Ihmels J, Barkai N Similarities and differences in genome-wide expression data of six organisms PLoS Biol. 2004 Jan;2(1).

[45] Kelley BP, Sharan R, Karp RM, Sittler T, Root DE, Stockwell BR, Ideker T. Conserved pathways within bacteria and yeast as revealed by global protein network alignment. Proc Natl Acad Sci U S A. 2003 Sep 30;100(20):11394-9.

[46] Kelley R, Ideker T. Systematic interpretation of genetic interactions using protein networks. Nat Biotechnol. 2005 May;23(5):561-6.

[47] Zhang LV, King OD, Wong SL, Goldberg DS, Tong AH, Lesage G, Andrews B, Bussey H, Boone C, Roth FP. Motifs, themes and thematic maps of an integrated Saccharomyces cerevisiae interaction network. J Biol. 2005;4(2):6.

[48] Sharan R, Ideker T. Modeling cellular machinery through biological network comparison. Nat Biotechnol. 2006 Apr;24(4):427-33.

[49] McCarroll SA, Murphy CT, Zou S, Pletcher SD, Chin CS, et al. (2004) Comparing genomic expression patterns across species identifies shared transcriptional profile in aging. Nat Genet 36: 197-204.

[50] Ihmels J, Bergmann S, Berman J, Barkai N. The differential clustering approach for comparative gene expression analysis: application to the Candida albicans transcription program. PLoS Genetics (2005) Vol. 1, No. 3, e39.

[51] Yvert G, Brem RB, Whittle J, Akey JM, Foss E, Smith EN, Mackelprang R, Kruglyak L (2003). Trans-acting regulatory variation in Saccharomyces cerevisiae and the role of transcription factors. Nat Genet 35: 57-64.

Studying evolutionary constraints on gene expression regulation *in silico*

Yonatan Bilu[1], Tomer Shlomi[2], Naama Barkai[1], Eytan Ruppin[2]
[1] Weizmann Institute of Science
Rehovot 76100, Israel
[2] Tel Aviv University
Tel-Aviv 69978, Israel

Abstract

Variation in gene expression levels on a genomic-scale has been detected among different strains, among closely related species and within populations of genetically identical cells. What are the driving forces which lead to expression divergence in some genes, and conserved expression in others? Here we employ Flux Balance Analysis (FBA) to address this question for metabolic genes. We consider the genome-scale metabolic model of *S. cerevisiae*, and its entire space of optimal and near-optimal flux distributions. We show that this space reveals underlying evolutionary constraints on expression regulation, as well as on the conservation of the underlying gene sequences. Genes which have a high range of optimal flux levels tend to display divergent expression levels among different yeast strains and species. This suggests that gene regulation has diverged in those parts of the metabolic network which are less constrained.

an extended version of this work was submitted to PLoS CB

Introduction

Recent comparative studies of genomic-scale gene expression levels have revealed substantial variation among different strains [1, 2], among closely related species [3], and even within a genetically identical population [4, 5, 6, 7]. Why do some of the genes monitored in these experiments manifest divergent expression values while the expression of others is constrained? We study this question from an evolutionary perspective in an in silico model of metabolism.

A key tool in studying metabolic networks is constraint-based modeling, which permits analysis of large-scale networks. Accurate prediction of dynamic metabolic activity requires kinetic models, but these rely on detailed information of the rates of enzyme activity, which is mostly unavailable, and are thus limited to small-scale networks. In contrast, constraint-based models use genome-scale networks to predict steady-state metabolic activity, regardless of specific enzyme kinetics. Stoichiometric, thermodynamic, flux capacity and possibly other constraints are used to limit the space of possible flux distributions attainable by the metabolic network. Flux Balance Analysis (FBA) [8, 9] is a specific constraint-based method which assumes that the network is regulated to maximize or minimize a certain cellular function, which is usually taken to be the organism's growth rate. FBA has been successfully used for predicting growth, uptake rates, by-product secretion, and growth following adaptive evolution, as well as other phenotypes [10, 11, 12, 13, 14].

The metabolic state predicted by FBA for a given growth media is not unique - in many cases there is a set of an infinite number of optimal solutions. Thus, we discuss here the optimal solution space, the space of all flux distributions leading to an optimal growth rate. On one hand, there are missing constraints in the model, and one line of research aims at reducing the solution space by adding biologically plausible constraints, e.g., by explicitly incorporating regulatory constraints in the model [15] and by looking for specific reactions for which new constraints may significantly reduce the size of the solution space [16]. On the other hand, even though there are still probably some missing constraints in FBA models, it has already been demonstrated that the solution space of such models does carry meaningful biological information [17, 18]. Alternative optimal steady-state flux solutions were shown to reflect redundant pathways [19], and sampling of the FBA solution space has been used, for example, to identify correlated fluxes in the mitochondrial metabolic network [18]. This point has been largely ignored in many FBA studies which, focusing on a variety of other research questions, have examined an arbitrary single optimal solution.

Here we suggest that the FBA solution space reflects multiple biologically meaningful metabolic states that are active in various conditions or under different evolutionary trajectories. We study the possibility that it reflects evolutionary constraints on expression and sequence of the associated genes. In particular, we focus on yeast metabolism using a recent genome-scale, fully compartmentalized model of *S. cerevisiae* [20], which is one of the largest and most comprehensive metabolic models available for a micro-organism. We show that by considering the entire FBA solution space we can identify constraints on the regulation of the associated genes, and thus predict which genes will have divergent expression levels among different yeast strains, and less conserved regulation among closely related species. These results show that the space of FBA solutions, which emerges from a complex interplay between the stoichiometric constraints, the uptake rates defining the growth medium, and the optimality assumption, is not just a technical consequence of our ignorance of additional constraints. Rather, this solution space sheds light on the evolution of metabolic regulation and of the metabolic network itself.

Results

Cellular metabolism is governed by various factors such as enzyme kinetics, allosteric control, transcriptional and post-transcriptional gene and protein regulation. Specifically, the effect of transcriptional regulation on cellular metabolism was previously studied based on gene expression measurements, small-scale flux measurements and large-scale flux predictions. Enzyme coding genes that form metabolic pathways were shown to be expressed 'just-in-time' when needed in bacteria [21]. A strong qualitative correspondence between gene expression and metabolic fluxes for various pathways was shown in both bacteria and yeast following environmental changes in yeast [22], and adaptive evolution in bacteria [23]. Previous studies have also shown that the expression patterns of enzyme coding genes are correlated with the flux patterns predicted by FBA: Schuster et al. and Famili et al. have shown that genes, associated with fluxes which are predicted to change together when shifting from one medium to another (e.g. in diauxic shift), are co-expressed under these conditions (but this was done on a small scale, esp. the former), while Reed and Palsson have shown the genes associated with fluxes that are correlated within the solution space also exhibit moderate levels of correlation in their expression [12, 24, 25].

Here we identify a more direct relation between expression and flux. We compared mRNA transcript numbers [26, 27] and protein levels [28] in rich media (YPD) to the predicted flux values when simulating YPD growth conditions (see Methods). As shown in Table , we find that the flux values show a significant correlation with the corresponding gene expression levels and with protein abundance data measured via GFP fluorescence. Isozymes were not included in this analysis, but their inclusion yields qualitatively similar results.

Under most simulated conditions, FBA has an infinite number of optimal solutions. As shown in Figure , some reactions display a broad range of values within the set of optimal solutions for glucose-rich condition, while others have an almost fixed value in all optimal solutions. A reaction displays a broad range of values when there are alternative pathways to the one it belongs to. For example, when simulating glucose-rich conditions, the reaction along the glycolysis pathways that is catalyzed by Fba1, which converts fructose 1,6 bisphosphate into glyceraldehyde-3-phosphate and dihydroxyacetone phosphate, can have a flux approximately equal to that of glucose intake, or, alternatively, be bypassed completely via the pentose phosphate pathway.

As flux values are significantly correlated with expression levels, we hypothesized that the range of possible optimal flux values for a given reaction reflects evolutionary constraints on the expression levels of its associated enzymes. Specifically, that the regulation of reactions which have an optimal fixed value is under strong selection to maintain their flux at the precise levels needed, while the regulation of reactions which may have a broad range of optimal values is under weaker selection.

To pursue this possibility we used flux variability analysis ([19, 29]): for each reaction we computed the maximal and minimal flux values attainable in the space of optimal flux distributions for growth conditions simulating YPD rich media. We define the YPD-flexibility score of a reaction as the difference between these maximal and minimal values. In addition, we performed a similar analysis, this time computing flux flexibility scores

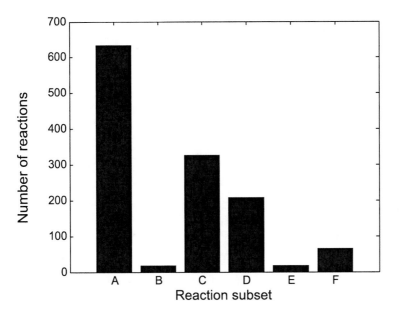

Figure 1: Distribution of reactions according their variability within the space of optimal solutions for YPD. (A) Reactions with flux equal to zero in all optimal solutions. (B) Reactions with the same, non-zero value in all optimal solutions. (C) Reactions with low variability among optimal solutions (flexbility score between 0 and 10^{-2}). (D) Reactions with medium variability among optimal solutions (flexbility score between 10^{-2} and $1/2$). (E) Reactions with high variability among optimal solutions (flexibility score at least $1/2$).

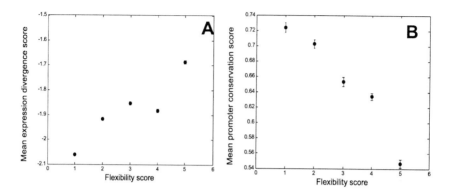

Figure 2: Correlation of YPD-flexibility score with biological measurements reflecting constraints on regulation: (A) Mean expression divergence (from Townsend et al.) as a function of the gene's flexibility score; (B) Mean promoter conservation score (based on Harbison et al.) as a function of the gene's. Genes are binned according to their flexibility score into five bins, such that each bin contains the same number of genes. Plot-points represent mean y-axis values over these bins. Error bars depict mean standard error.

across 1000 randomly generated growth media. In both cases, the flexibility score of a gene has been taken as the maximal score among the flexibility scores of the reactions it is associated with. Interestingly, we found that the YPD-flexibility score is very similar to the flexibility scores found across the random media (mean Spearman rank correlation = 0.92), and focus on it in the sequel.

To test the hypothesis that genes with smaller YPD-flexibility scores have tighter regulation, we studied the correlation between the genes' YPD-flexibility scores, the conservation of their promoters and their expression patterns. We used a score computed by Townsend, Cavalieri et al. [2], which measured expression patterns for four yeast strains, obtaining an expression divergence score for each gene. Harbison, Gordon et al. [30] constructed a comprehensive map of S. cerevisiae transcription factor binding sites, and for each site reported the number of yeast species (from among *S. paradoxus, S. mikatae* and *S. bayanus*) in which it is conserved. Based on this data, we assigned to each gene a promoter conservation score (see Methods). As shown in Figures a and b, both scores show significant correlations with the YPD-flexibility score; the Spearman rank correlation with expression divergence is 0.18 (p-value $< 10^{-4}$, 469 genes), and is -0.18 with promoter conservation (p-value $= 10^{-3}$, 330 genes). Although these significant correlations are of moderate magnitude, they are robust, and taken together they support the hypothesis that low flexibility scores are associated with conserved regulation.

A likely mechanism underlying these findings is that genes with high flexibility scores take part in pathways which have alternative ones for optimal growth. In some conditions one pathway is used, while in others the alternative is taken, leading to a variable expression pattern. This is reflected not only in the model, but also in comparison with lethality assays. Of the 50 genes with highest flexibility score only 6% are essential for growth in

YPD, whereas of the 89 which have zero flexibility score, 17% are essential.

Discussion

What do the metabolic flux distributions composing the FBA solution space represent? While some of them may be superfluous, arising from missing constraints, this study shows that as a whole, they are biologically meaningful. Three different possibilities for their interpretation suggest themselves:

1. *Effect of exogenous factors*: In addition to the growth medium simulated by the model, the metabolic behavior of an organism is also influenced by exogenous factors which are beyond the model's scope. Hence, it is plausible that the stoichiometric constraints allow for the existence of a variety of different metabolic states which are needed for growth in the given medium under the various different external conditions that the organism may encounter. The solution space represents the union of these different conditions.

2. *Alternative evolutionary pathways*: It was suggested that the FBA solution space may contain alternative flux distributions that are attainable through different evolutionary paths, as was experimentally shown in an adaptive evolution experiment in E. coli [19, 23, 31, 32]. These experiments showed that although evolutionary endpoints may converge with respect to the selection pressure for high growth rate, the underlying metabolic states, characterized by measured metabolic fluxes and gene expression may significantly diverge.

3. *Heterogeneity within a population*: Previous studies suggest that the predicted variability in metabolic states may represent heterogeneous metabolic behaviors of individuals within a cell population [17, 19]. It is possible that the multiple metabolic states composing the space of optimal solutions represents this heterogeneity. This possibility is especially appealing in light of recent measurements of gene expression at the single cell level, showing high variability in expression among cells taken from the same culture [4, 5, 6, 7].

These interpretations are not exclusive, and the FBA solution space is likely to represent an interplay between them. Here we focus on evidence supporting interpretation 2. We see that that reactions which display a range of values within the space of optimal solutions tend to be associated with genes whose regulation is less conserved. In other words, where optimality allows for different evolutionary paths to be taken, the data suggests that they are indeed taken. This finding is in agreement with the previous experiments of bacterial adaptive evolution, showing that adaptive mechanisms evolved in the transcriptional regulatory network that governs their metabolic state [32].

In summary, we have shown that genes with a high potential flux range have indeed less constraints on their regulation. These results emerge when studying the FBA solution space as a whole, clearly showing that it carries meaningful biologically information.

Methods

Metabolic network model

The S. cerevisiae metabolic model used here is by [20]. It includes 1149 reactions, associated with 734 genes. A reaction is considered active in a given flux distribution, if its associated flux is non zero. Of the 1149 reactions, 268 were identified to be essentially zero (absolute value at most 10^{-10}) in all feasible (not necessarily optimal) flux distributions satisfying the stoichiometric constraints simulating YPD. Since these reactions are not relevant for growth in YPD they were omitted when computing the YPD-flexibility score and YPD-activity score.

In addition to the 1149 internal reactions, we added to the model 116 uptake/excretion reactions, for each of the metabolites listed as "extra-cellular" in the basic model.

The optimal solution space is defined as the set of all flux distributions which obey the stoichiometric and thermodynamic constraints, lead to a maximal growth rate, and minimize the sum of (absolute values of) the reactions. The latter constraint, following [33], aims to avoid futile flux cycles in the network which violate the laws of thermodynamics.

Datasets used

The mRNA transcript numbers was taken from [26] and [27], which list values for 679 and 728 of the genes included in model, respectively.

Protein abundance, as measured through GFP fluorescence, was taken from [28, 34, 35], and was available for 475 of the encoding genes in the model.

Expression divergence measures were taken from the supplementary results of [2], 581 of which are for genes which appear in the model.

Simulating random growth media

Random growth media were generated by setting limiting values to the uptake reactions independently at random. With probability p, the maximal uptake rate was set to 0 - i.e. only excretion was allowed. Otherwise, uptake rate was limited to a value chosen uniformly at random between 0 and 1.

The values p=0.5 and p=0.95 were tested, simulating rich and poor media, respectively. Eight of the uptake rates were taken positive in all media, to ensure viability (for water, ammonium, phosphate, sulfate, oxygen, sodium, potassium and carbon dioxide).

A similar sampling method was used in [36].

Gene flexibility score

The flexibility score of a reaction in a given medium is the difference between the maximal flux that can flow through it in an optimal flux distribution, and the minimal one. This is computed using flux variability analysis [19].

The flexibility score of a gene in a given medium is the maximum of the flexibility scores for the reactions it is associated with. The rationale for using the maximum value is that this value is the one that most constrains the required enzyme quantity for obtaining optimal flux. That is, if a protein is associated with several reactions, for its expression level to comply with all optimal flux values, it most complies with the highest one.

For isozymes the definition is slightly more complicated. Suppose a reaction R can attain values between x and y, and hence its flexibility score is $y - x$. If a single gene is associated with this reaction, then, as defined above, the reaction's contribution towards the gene's flexibility

Table 1: Spearman rank correlation between flux level and mRNA/protein abundance. Values are based on flux levels from 1000 randomly sampled optimal flux distributions.

Source	Data type	Mean correlation	Standard deviation	Mean p-value	Number of genes
26	mRNA number	0.35	0.006	$2x10^{-11}$	343
27	mRNA number	0.37	0.006	10^{-12}	356
28	Protein level	0.22	0.01	$4x10^{-4}$	259

score is simply $y - x$, since we think as this range of fluxes as defining the flexibility of the gene's expression. However, if there are several isozymes associated with the reaction, then their flexibility can, potentially, be larger.

If $x > 0$ then the values an associated isozyme I can attain are from 0 to y, since if the expression level of I is smaller than x, the other isozymes can compensate for it, putting the total reaction rate with the optimal range of $[x, y]$. Hence, in this case the contribution of the reaction towards the genes flexibility is y, rather than $y - x$. Similarly, if $y < 0$, an isozyme could attain any value in the range $[x, 0]$, and hence the contribution of the reaction towards the gene's flexibility would by $|x|$. Taken together, we define contribution of a reaction attaining flux values in the range $[x, y]$ towards the flexibility score of its associated isozymes as the maximum among $y - x$, y, and $|x|$.

Promoter conservation score

Transcription factors' binding sites were considered as regulating a gene if they appear in the map of [30] within 500bp of the gene's translation start site.

Each binding site received a score of 1, if it is conserved in all three species from among *S. paradoxus*, *S. mikatae* and *S. bayanus* ; a score of 0.5 if it conserved in two of them, and a score of 0 otherwise. The promoter conservation score for a gene is the mean score for the binding sites associated with it.

Data was available for 382 genes.

Bibliography

[1] Fay JC, McCullough HL, Sniegowski PD, Eisen MB (2004) Population genetic variation in gene expression is associated with phenotypic variation in Saccharomyces cerevisiae. Genome Biol 5: R26.

[2] Townsend JP, Cavalieri D, Hartl DL (2003) Population genetic variation in genome-wide gene expression. Mol Biol Evol 20: 955-963.

[3] Wittkopp PJ, Haerum BK, Clark AG (2004) Evolutionary changes in cis and trans gene regulation. Nature 430: 85-88.

[4] Braun E, Brenner N (2004) Transient responses and adaptation to steady state in a eukaryotic gene regulation system. Phys Biol: 67-76.

[5] de Atauri P, Orrell D, Ramsey S, Bolouri H (2005) Is the regulation of galactose 1-phosphate tuned against gene expression noise? Biochem J 387: 77-84.

[6] Raser JM, O'Shea EK (2004) Control of stochasticity in eukaryotic gene expression. Science 304: 1811-1814.

[7] Swain PS, Elowitz MB, Siggia ED (2002) Intrinsic and extrinsic contributions to stochasticity in gene expression. Proc Natl Acad Sci U S A 99: 12795-12800.

[8] Fell DA, Small JR (1986) Fat synthesis in adipose tissue. An examination of stoichiometric constraints. Biochem J 238: 781-786.

[9] Kauffman KJ, Prakash P, Edwards JS (2003) Advances in flux balance analysis. Curr Opin Biotechnol 14: 491-496.

[10] Edwards JS, Ibarra RU, Palsson BO (2001) In silico predictions of Escherichia coli metabolic capabilities are consistent with experimental data. Nat Biotechnol 19: 125-130.

[11] Edwards JS, Palsson BO (2000) The Escherichia coli MG1655 in silico metabolic genotype: its definition, characteristics, and capabilities. Proc Natl Acad Sci U S A 97: 5528-5533.

[12] Famili I, Forster J, Nielsen J, Palsson BO (2003) Saccharomyces cerevisiae phenotypes can be predicted by using constraint-based analysis of a genome-scale reconstructed metabolic network. Proc Natl Acad Sci U S A 100: 13134-13139.

[13] Forster J, Famili I, Palsson BO, Nielsen J (2003) Large-scale evaluation of in silico gene deletions in Saccharomyces cerevisiae. Omics 7: 193-202.

[14] Ibarra RU, Edwards JS, Palsson BO (2002) Escherichia coli K-12 undergoes adaptive evolution to achieve in silico predicted optimal growth. Nature 420: 186-189.

[15] Covert MW, Knight EM, Reed JL, Herrgard MJ, Palsson BO (2004) Integrating high-throughput and computational data elucidates bacterial networks. Nature 429: 92-96.

[16] Wiback SJ, Famili I, Greenberg HJ, Palsson BO (2004) Monte Carlo sampling can be used to determine the size and shape of the steady-state flux space. J Theor Biol 228: 437-447.

[17] Price ND, Reed JL, Papin JA, Wiback SJ, Palsson BO (2003) Network-based analysis of metabolic regulation in the human red blood cell. J Theor Biol 225: 185-194.

[18] Thiele I, Price ND, Vo TD, Palsson BO (2005) Candidate metabolic network states in human mitochondria. Impact of diabetes, ischemia, and diet. J Biol Chem 280: 11683-11695.

[19] Mahadevan R, Schilling CH (2003) The effects of alternate optimal solutions in constraint-based genome-scale metabolic models. Metab Eng 5: 264-276.

[20] Duarte NC, Herrgard MJ, Palsson BO (2004) Reconstruction and validation of Saccharomyces cerevisiae iND750, a fully compartmentalized genome-scale metabolic model. Genome Res 14: 1298-1309.

[21] Zaslaver A, Mayo AE, Rosenberg R, Bashkin P, Sberro H, et al. (2004) Just-in-time transcription program in metabolic pathways. Nat Genet 36: 486-491.

[22] Daran-Lapujade P, Jansen ML, Daran JM, van Gulik W, de Winde JH, et al. (2004) Role of transcriptional regulation in controlling fluxes in central carbon metabolism of Saccharomyces cerevisiae. A chemostat culture study. J Biol Chem 279: 9125-9138.

[23] Fong SS, Palsson BO (2004) Metabolic gene-deletion strains of Escherichia coli evolve to computationally predicted growth phenotypes. Nat Genet 36: 1056-1058.

[24] Reed JL, Palsson BO (2004) Genome-scale in silico models of E. coli have multiple equivalent phenotypic states: assessment of correlated reaction subsets that comprise network states. Genome Res 14: 1797-1805.

[25] Schuster S, Klamt S, Weckwerth S, Moldenhauer F, Pfeiffer T (2002) Use of network analysis of metabolic systems in bioengineering. Bioprocess and Biosystems Engineering 24: 363 - 372.

[26] Holstege FC, Jennings EG, Wyrick JJ, Lee TI, Hengartner CJ, et al. (1998) Dissecting the regulatory circuitry of a eukaryotic genome. Cell 95: 717-728.

[27] Velculescu VE, Zhang L, Zhou W, Vogelstein J, Basrai MA, et al. (1997) Characterization of the yeast transcriptome. Cell 88: 243-251.

[28] Huh WK, Falvo JV, Gerke LC, Carroll AS, Howson RW, et al. (2003) Global analysis of protein localization in budding yeast. Nature 425: 686-691.

[29] Reed JL, Palsson BO (2003) Thirteen years of building constraint-based in silico models of Escherichia coli. J Bacteriol 185: 2692-2699.

[30] Harbison CT, Gordon DB, Lee TI, Rinaldi NJ, Macisaac KD, et al. (2004) Transcriptional regulatory code of a eukaryotic genome. Nature 431: 99-104.

[31] Fong SS, Nanchen A, Palsson BO, Sauer U (2006) Latent pathway activation and increased pathway capacity enable Escherichia coli adaptation to loss of key metabolic enzymes. J Biol Chem 281: 8024-8033.

[32] Fong SS, Joyce AR, Palsson BO (2005) Parallel adaptive evolution cultures of Escherichia coli lead to convergent growth phenotypes with different gene expression states. Genome Res 15: 1365-1372.

[33] Kuepfer L, Sauer U, Blank LM (2005) Metabolic functions of duplicate genes in Saccharomyces cerevisiae. Genome Res 15: 1421-1430.

[34] Ghaemmaghami S, Huh WK, Bower K, Howson RW, Belle A, et al. (2003) Global analysis of protein expression in yeast. Nature 425: 737-741.

[35] Ghaemmaghami S, Huh WK, Bower K, Howson RW, Belle A, et al. (2003) Yeast GFP-fusion Localization Database.

[36] Almaas E, Oltvai ZN, Barabasi AL (2005) The Activity Reaction Core and Plasticity of Metabolic Networks. PLoS Comput Biol 1: e68.

Metabolic Networks

Vertical Genomics

Jildau Bouwman, A. Canelas, Sergio Rossell, Karen van Eunen
Hans V. Westerhoff, and Barbara M. Bakker
Molecular Cell Physiology and Molecular Cell Biology
1081HV, Vrije Universiteit, Amsterdam, The Netherlands

Abstract
In the Vertical genomics project we would like to answer the question: How does the concerted action of metabolism and gene expression regulate a metabolic flux? Cell function depends critically on the interaction between various types of macromolecular processes. These include transcription, translation, metabolism and transport. Most often studies are focused on a single (horizontal) level in the cell's hierarchy, for instance at the transcription of all genes. However, in order to understand the functioning of the cell, it is also important to identify how the cell regulates its function at all levels of organization simultaneously. This may be different for various conditions and enzymes. Vertical Genomics looks at all levels for a subset of genes.

Introduction

When cells are brought from one condition to another, the cell has to adapt. In the new situation, the cell might for instance need more energy. Therefore, the pathway involved in the breakdown of sugars needs to be more active (more flux through the pathway). Changing the activity of a pathway might be a result of a change in intermediates (changes at the metabolic level), modifications of enzymes or by an increase in the amount of enzymes. The cell can increase the amount of enzymes by increasing the gene expression. Gene expression includes transcription (the step from DNA to RNA) and translation (the RNA is used to make proteins). All these steps, but also the degradation of RNA and protein can be regulated by the cell. Although the mechanisms of all those processes are well known, which of the steps are in practice used to regulate the flux through a pathway is hardly known. We hypothesize that cells adapt by simultaneous regulation at all levels (Fig. 1). Methods are now available to monitor all the components. Glycolysis in yeast is a good model system to unravel at what level yeast regulates its flux, as it is one of the few pathways for which the kinetic properties of the enzymes are known sufficiently to calculate the flux from the enzyme activities and yeast can be brought under the well-defined steady-state and transient conditions.

In the 'Vertical Genomics' project steady-state yeast cultures are perturbed in different ways, including shifting from aerobic to anaerobic conditions, a temperature-shift, nitrogen starvation, osmotic shock and by shifting from respiratory to fermentative growth. Using regulation analysis [1] we will quantify to what extent these changes are caused by changes in transcription, transla-

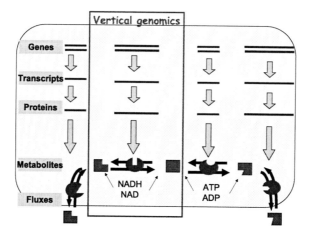

Figure 1:

tion and/or metabolism. RNA concentrations, protein concentration, metabolites concentrations, enzyme activities and fluxes are measured quantitatively. In addition, we will try to quantify to what extend production and degradation of RNA and protein are involved in regulating cell function. Standardization is a key issue in this project, because cultivation as well as sample analysis is done in 6 different labs. Standardization will make it possible to make one model of the whole vertical cascade (Fig. 2) with data from different labs. We standardized the protocols we use and additionally we all start from the same standard culture. We would like to make a distinction between the role of transcription and degradation of RNA in adjusting cell function. However, we are only able to measure RNA levels, which are affected by both processes. Blocking one of the processes makes it possible to measure the other. Transcription can easily be blocked by inhibitors. However, if we block transcription we cannot normalize to the currently used internal standard, because also the RNA levels of our control will be affected by treating the cell with a transcription blocker. To overcome these practical problems, we are currently setting up a method to measure absolute numbers of RNA copies per cell.

Results

The steady-state properties of the standard culture from different labs are similar and the amount of RNA for several proteins is comparable in the standard cultures of different labs. This shows that we are able to standardize our culturing conditions. One of the conditions studied in the project is the glucose pulse. The cells are first grown under the standard conditions (a glucose-limited chemostat) and then a large amount of glucose is added to the culture. Under these conditions the cell adapts by reducing the RNA levels for several enzymes in glycolysis rapidly. Some of the RNA species are reduced within 3 minutes. An osmotic shock (by adding 1 M of sorbitol) shows a reversed picture. If we look at the RNA levels of the glycerol pathway we show that the RNA levels rapidly increase after the shock. If yeast is perturbed to a condition of nitrogen starvation,

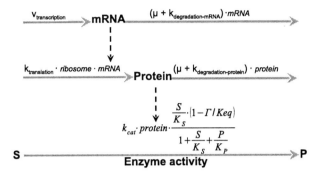

Figure 2:

we show that gene expression is not the only process that regulates the flux though glycolysis, but also the changes occur at the metabolic level.

Conclusions

We are able to standardize our culturing conditions to that extend that the standard culture in different labs is similar. Therefore, we can use the data from different labs to build one model. We show that yeast uses several gene expression levels to change the amount of enzymes in glycolysis and in this way adjust the flux. We see an increase and a decrease in mRNA levels. Under other conditions and for different enzymes the flux is adjusted by changes at the metabolic level. Also a combined effect of the changes at the metabolic level and in the gene expression are seen.

Bibliography

[1] Rossell, S., van der Weijden, C.C., Lindenbergh, A., van Tuijl, A., Francke, C., Bakker, B.M. and Westerhoff, H.V. Unraveling the complexity of flux regulation: a new method demonstrated for nutrient starvation in Saccharomyces cerevisiae. Proc Natl Acad Sci U S A. 103, 2166-71 (2006)

An Overview of Computational Approaches to Metabolic Networks

Ursula Kummer
EML Research
69118 Heidelberg, Germany

Abstract

Metabolic networks, consisting of many different biochemical reactions form the basis of cellular function. Diseases often originate in a malfunctioning of the metabolic network. Therefore, whole scientific fields, e.g. biochemistry are concerned with the elucidation of this function and the underlying mechanism. Due to the complexity of this task computational approaches have become more and more important. The most important ones are reviewed in this overview.

Introduction to Metabolic Networks

Thousands of different biochemical reactions run in each living cell at any given moment. These reactions are connected by means of mass transfer (e.g. they form consecutive or branched chains of reactions) as well as by information transfer (e.g. one compound of one reaction modifies the speed of another reaction). Thus, they are the constituents of a huge, highly connected and complex network.

The central importance of this network for every living organism has led to the fact that the elucidation of its function and its underlying mechanisms is probably the biggest scientific effort in terms of personnel and resources right now. Biochemistry, physiology and cell biology are only a few of the many research fields that are busy with this task. Lots of technical advances have been made in the last few years which allow e.g. high-throughput sampling of data and/or high-resolution measurements of specific compounds in the cell. Thus, it is now possible (see e.g. subchapter data in this chapter) to gather information about the concentration of different metabolic species, proteins and other compounds at different points in time.

However, this is not sufficient to really get an understanding of what is going on. The sheer complexity of the task asks for computational support and therefore, computational biochemistry (or computational systems biology) has become a very important part of the biosciences recently. Different layers of abstraction are addressed by different computational techniques and an overview of the most important ones (subjectively selected by the author) are presented in the remainder of this chapter.

Studying Network Topologies

The most abundant information about biochemical reactions available is their mere presence or absence in an organism. Therefore, a relatively easily accessible topic is the study of static networks with no weights on their edges/nodes (see also introduction of this book). In most cases, however, the connectivity which arises from information transfer, namely when compounds act as modifiers of other reactions, are left out.

Thus, there are quite a number of studies which depict the mass transfer part of a metabolic network in which nodes represent metabolites and edges represent reactions. These studies look at the connectivity and diverse topological measures. Table 1 represents the results of such a study by Fell and collaborators [1] for *E.coli*. In addition, identification and analyses of motifs within the static network are a popular topic (for reviews see [2]), citegrigo. Implications for the evolution of the networks are discussed by these and other authors, but not really addressed computationally.

Rank by degree	Connectivity	Rank by mean path length	Mean path length
glutamate	51	glutamate	2.46
pyruvate	29	pyruvate	2.59
CoA	29	CoA	2.69
2–oxoglutarate	27	glutamine	2.77
glutamine	22	acetyl CoA	2.86
aspartate	20	oxoisovalerate	2.88
acetyl CoA	17	aspartate	2.91
phosphoribosylPP	16	2–oxoglutarate	2.99
tetrahydrofolate	15	phosphoribosylPP	3.10
succinate	14	anthranilate	3.10
3–phosphoglycerate	13	chorismate	3.13
serine	13	valine	3.14
oxoisovalerate	12	3–phosphoglycerate	3.15

Table 1: Kindly provided by Fell et al. [1]. Thirteen key metabolites of *E. coli* metabolism. Metabolites with connectivity significantly higher than the mean metabolite degree are shown. For comparison, the thirteen metabolites with the shortest mean path lengths are shown.

On the other hand static topological networks with weights have been studied quite extensively. Only few examples deal with weights attached to the nodes e.g. [4] inferring metabolic pathways from experimental data. The vast majority deals with weights attached to the edges, in most cases representing the stoichiometry of the respective reaction. Thus, analysis of these weighted networks allows the determination of subnetworks that can proceed by themselves running into a steady state. These subnetworks have been calles extreme currents, elementary modes or extreme pathways with only slightly different definitions ([5] for a review). The study of any of these can be very useful, e.g. to predict the consequences of the loss or the addition of an edge, meaning the malfunctioning or addition of an enzyme to a biochemical system.

Computational analysis of evolutionary static networks with weights are not existing (to the best of my knowledge).

Studying Dynamic Networks

Even though our knowledge about the parameters governing the velocity of biochemical reactions *in vivo* is rather limited, the by far dominating computational field in the study of metabolic networks is the modelling, simulation and analysis of dynamic networks. The reason for that is likely that these models, if set up properly, are doubtless the strongest w.r.t. supporting our understanding of the biochemistry and w.r.t. predictions. The latter are of course the ultimate goal of any serious modeling effort.

Dynamic networks are modeled with various computational approaches. Most commonly, ordinary and partial differential equations are used (introductory text [6]). Here, metabolites, the nodes in our networks are the variables and reactions are presented by kinetic terms that add up to the right-hand side of the equations. These kinetic terms typically depend on some of the variables of the system. A typical example is given below describing calcium signal transduction in mammalian hepatocytes [7]:

$$J_{ER,ch} = k_{10} Ca_{cyt} PLC \frac{Ca_{ER}}{K_{11} + Ca_{ER}},$$

$$\frac{dCa_{cyt}}{dt} = J_{ER,ch} - J_{ER,pump} + J_{in} - J_{out}, \qquad J_{ER,pump} = k_{16} \frac{Ca_{cyt}}{K_{17} + Ca_{cyt}},$$

$$\frac{dCa_{ER}}{dt} = J_{ER,pump} - J_{ER,ch}, \qquad J_{in} = k_{12} PLC + k_{13} G_{\alpha},$$

$$\frac{dPLC}{dt} = J_{PLC,act} - J_{PLC,inact}, \qquad J_{out} = k_{14} \frac{Ca_{cyt}}{K_{15} + Ca_{cyt}},$$

$$\frac{dG_{\alpha}}{dt} = J_{G_{\alpha},act} - J_{G_{\alpha},inact}, \qquad J_{PLC,act} = k_{7} G_{\alpha},$$

In addition to the simulation of these systems, steady-state, stability and sensitivity are among the most common computational analysis used. Many biochemical systems exhibit pronounced nonlinear dynamics. Therefore, many tools from the theory of nonlinear dynamics are also frequently applied, e.g. continuation as shown in Fig. 1.

Apart from the common use of differential equations, other approaches to dynamical networks have been used recently. Among these, cellular automata, (recent example see [8]), Petri-Nets (for a comparison with conventional approaches see [9]), process algebra, e.g. Pi-calculus [10], and other discrete event simulations, e.g. [11] are the most popular. However, it would exceed the scope of this short overview to go into these in any detail.

Within UniNet, our work focuses on the prediction of dynamic properties (e.g. stability) of static networks with weights. This is described in the following research paper.

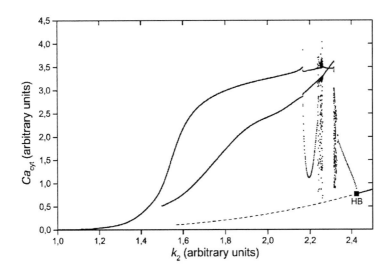

Figure 1: Bifurcation diagram of a model describing calcium signal transduction in hepatocytes.

Acknowledgements

Apart from the EU funding for UniNet, I would like to thank the Klaus Tschira Foundation for financing.

Bibliography

[1] Fell, D.A. and A. Wagner. The Small world inside large metabolic networks. In: R. Gauges, C. van Gend and U. Kummer (eds), 2nd Workshop on Computation of Biochemical Pathways and Genetic Networks. Logos Verlag, Berlin, pp. 11-19, 2001. ISBN 3-89722-648-0.

[2] Xia Y., Yu H., Jansen R., Seringhaus M., Baxter S., Greenbaum D., Zhao H. and M. Gerstein. Analyzing cellular biochemistry in terms of molecular networks. Annu. Rev. Biochem. 2004, 73:1051-87.

[3] Grigorov M.G. Global properties of biological networks. Drug Discov. Today. 2005, 10:365-72.

[4] Croes, D., Couche, F., Wodak, S.J. and J. van Helden. Inferring meaningful pathways in weighted metabolic networks. J. Mol. Biol. 2006, 356:222-36.

[5] Schilling C.H., Schuster S., Palsson B.O. and R. Heinrich. Metabolic pathway analysis: basic concepts and scientific applications in the post-genomic era. Biotechnol. Prog. 1999, 15:296-303.

[6] R. Heinrich and S. Schuster. The Regulation of Cellular Systems. Kluwer Academic Publishers, Amsterdam. ISBN 0-41203-261-9.

[7] Grubelnik V., Larsen A.Z., Kummer U., Olsen L.F. and M. Marhl. Mitochondria regulate the amplitude of simple and complex calcium oscillations. Biophys Chem. 2001, 94:59-74.

[8] Wishart D.S., Yang R., Arndt D., Tang P. and J. Cruz. Dynamic cellular automata: an alternative approach to cellular simulation. In Silico Biol. 2005, 5:139-61.

[9] Zevedei-Oancea I. and S. Schuster S. Topological analysis of metabolic networks based on Petri net theory. In Silico Biol. 2003, 3:323-45.

[10] Regev A., Silverman W. and E. Shapiro. Representation and simulation of biochemical processes using the pi-calculus process algebra. Pac. Symp. Biocomput. 2001, 459-70.

[11] Degenring D., Rohl M. and A.M. Uhrmacher. Discrete event, multi-level simulation of metabolite channeling. Biosystems. 2004, 75:29-41.

Stability analysis of metabolic networks

Iulian Stoleriu
EML Research, 69118 Heidelberg, Germany

Abstract
Metabolic networks are complex systems of interconnected biochemical reactions. Mathematically, such a complex network can be conveniently represented by a set of ordinary differential equations. Mathematical models for metabolism aim to improve the understanding of metabolic regulation by quantifying essential aspects of a metabolic system. Because of the large number of species involved in these networks and the fact that many parameters are not *a priori* known, some aspects of the entire network (e.g., the stability) are difficult to analyse. Therefore, mechanisms for network complexity reduction are required. In this paper, we will employ a reduction mechanism that uses extreme currents, i.e. we split the whole network into smaller subnetworks, which are easier to handle and analyse. We will give here some stability conditions, which may be expressed in inequalities among the rate parameters in the network.

Introduction

As biochemical networks grow in sizes, the reduction of their complexity becomes a very important issue. The main problems we face when studying large-scale systems of biochemical reactions are: the high number of kinetic parameters, most of them unknown *a priori*, the high dimensionality of the system, which makes the numerical simulations very demanding, and multiple time scales in the system, leading to stiff equation systems. Due to the high number of parameters in the system and the difficulty in finding the steady states, the linear stability analysis for high dimensional reaction networks appears to be a difficult task. We would like to be able to say something about the stability of these networks without having any *a priori* information about the kinetic parameters of the system. There would be also nice if one can find relations among kinetic parameters, which may drive the system towards a stable steady state, or relations for which one can be sure that the steady states will not be stable.

In this paper, we present a method which can be used in the stability analysis for the steady states of a general biochemical network. The method uses the stoichiometry information of the network and it is based on the split of the whole system into sub-systems, each such sub-system being generated by an extreme current. One can thus study the stability of each sub-network in part and find which extreme current may be responsable for instabilities in the whole network. We show here some results which may lead to inequalities among kinetic parameters in the system, from which one can decide upon stability or instability of the steady state of the entire network.

Stoichiometric network analysis

Stoichiometric network analysis is a general approach to study the qualitative dynamics of chemical networks, which has the advantage that the kinetic parameters are not needed in the analysis. For a good insight into the Stoichiometric Network Analysis, one can consult [2]. The time behaviour of an arbitrary metabolic network, consisting of m metabolites and r reactions, can be described by a system of differential equations, written in the following form:

$$\frac{dx(t)}{dt} = N \cdot v(x(t), k), \quad \text{for } t \geq 0,$$

where $x = (x_i)_{i=\overline{1,m}}$ is the vector having metabolites as components, $v = (v_j)_{j=\overline{1,r}}$ is velocity (reaction rates) vector, the vector k contains the kinetic parameters and $N \in \mathcal{M}_{m \times r}$ is the stoichiometric matrix, that is the matrix which contains in its jth column the stoichiometric coefficients for the jth reaction. The reaction rates v_j are usually monomials, such as $k_j \prod_i x_i^{\kappa_{ij}}$, where the order of kinetics κ_{ij} form an $m \times r$ matrix, denoted by κ, and called the kinetic matrix.

If a network contains conserved moieties, then these will correspond to linearly dependent rows in the stoichiometric matrix. It is then safe to remove dependent species from the stoichiometric matrix, as they will give rise to zero eigenvalues when studying linear stability of the system. Also, reducing the stoichiometry matrix will simplify both, the computation and the analysis. We will call a *reduced stoichiometric matrix*, the stoichiometric matrix which is obtained after deleting the rows corresponding to the dependent species in the network (due to conservation constraints). Let us denote by m_0 the dimension of the left null-space of N. If $m_0 = m$, then there will be no matrix reduction. If $m_0 < m$, then after a re-ordering of the rows, such that the first m_0 rows are linearly independent, we construct the reduced stochiometric matrix, N_r, by deleting the last $m - m_0$ rows from N. We now consider the vector of independent species, x_r, which are those who correspond to the rows in the matrix N_r, and write de reduced system as

$$\frac{dx_r}{dt} = N_r \cdot v(x, k), \quad \text{for } t \geq 0. \tag{1}$$

From now on we will be working with the reduced stoichiometric matrix N_r, instead of N, and the reduced kinetic matrix κ_r, instead of κ. The reduced matrix κ_r is obtained from κ exactly as N_r is deduced from N.

If the order of the matrix N_r is very high, then linear stability analysis will be a difficult task to accomplish. One needs thus to reduce the complexity of the system. The reduction method we will use in the following is via the consideration of extreme currents. By starting from the steady states set, one can the reduce the whole network into subnetworks generated by the extreme currents. One can study firstly the stability of these subnetworks and then conclude how the stability properties of them can influence the stability properties of the whole network.

Complexity reduction via extreme currents

For biochemical systems, the analysis of steady states is very important. The steady states are given by the system

$$N_r \cdot v(x, k) = 0, \quad v(x, k) \geq 0, \tag{2}$$

thus the feasible steady states are confined to the cone $\mathcal{K}_v = \{v \in \mathbb{R}^r; \ ker \, N \bigcap \mathbb{R}_+^r\}$. As shown in [1], $\mathcal{K}_v$ is spanned by a finite number of generating vectors, E_i, $i = \overline{1, q}$. They are called *extreme currents* and have the following properties: the set $\{E_i\}_i$ is unique, $N \cdot E_i = 0$, $\forall i = \overline{1, q}$ and any element v of $\mathcal{K}_v$ can be generated by a non-negative linear combination of the elements of this set, that is $\mathcal{K}_v = \{v \in \mathbb{R}^r; \ v(j) = \sum_{i=1}^{q} j_i E_i, \ j_i \geq 0\}$.

As shown in [1] and [4], the main advantage of this method is the possibility of re-writing the Jacobian of the system (1) in the following useful form:

$$J \ = \ N_r \cdot diag \ \left(\sum_{i=1}^{q} j_i E_i\right) \cdot \kappa_r^\tau \cdot diag \ (x_0^{-1}) \tag{3}$$

$$= \ \sum_{i=1}^{q} j_i \, N_r \cdot diag \ (E_i) \cdot \kappa_r^\tau \cdot diag \ (x_0^{-1}), \tag{4}$$

where x_0 is the steady states vector. Note that in these expressions of the Jacobian the kinetic rates k_i enter in the above formulae only through x_0 .

Because the mapping $x \rightarrow v(x, k)$ is not surjective (that is $\mathcal{K}_v$ is too wide for the steady state space), new restrictions on the v_i's are needed. In [3] and [4], the reaction rates v_i, which are some monomials, are restricted further to a variety of a toric ideal generated by a family of binomials of the form $v_i - v_i(x, k)$.

More explicitly, the reaction rates v_i are monomials and represent basis elements for a polynomial ideal. If we consider the ideal

$$I \ = \ \{f; \ f = \sum_{i=1}^{r} \zeta_i \, [v_i - v_i(x, k)] \, , \ v_i(x, k)\text{- reaction rates}\},$$

then the reaction rates lie in the space

$$V(I) \ = \ \{x \in \mathbb{C}^m; \ f(x) = 0, \ \forall f \in I\}.$$

We change the basis in the ideal I and use instead the Gröbner basis, based on the lexicographic order in monomials. The variety generated with the newly obtained binomials is a deformed toric ideal, which we denote by $I^{def, tor}$. Thus, the reaction rates will finally lie in the intersection of the polyhedral cone $\mathcal{K}_v$ and the variety generated by the deformed toric ideal, that is

$$v_i \in \mathcal{K}_v \bigcap V(I^{def, tor}), \ \forall i.$$

Stability analysis

We are now interested in studying the stability of the steady state of (5) and of the subnetworks generated by the extreme currents E_i. We start we the following definitions:

Definition 1: We call *essential eigenvalue* an eigenvalue of the matrix obtained from the restricted Jacobian matrix by deleting the rows corresponding to the non-essential species, i.e. species that do not appear in the reactions corresponding to the extreme current of interest.

Definition 2: We say that the steady state of (5) is *stable* if all essential eigenvalues of Jacobian matrix have negative real part. Moreover, we say that the subnetwork generated by E_i is *stable* if all essential eigenvalues of Jacobian matrix corresponding to E_i, $J(E_i)$, have negative real part.

Definition 3: We say that the subnetwork generated by E_i is *mixing stable* if $J(E_i) + J(E_i)^\tau$ has only essential eigenvalues with negative real part.

We interpret the mixing stability as the property of an extreme current to give rise of no instabilities when mixed with other extreme currents. As one can easily check, mixing stability implies stability in the above sens.

We have seen above that the Jacobian matrix in the v_i coordinates has the form given by (3). Thus, linear stability of the system (2) is reduced to the stability of the system

$$x' = A \cdot H \cdot x, \tag{5}$$

where $A = N_r \cdot diag \left(\sum_{i=1}^q j_i E_i \right) \cdot \kappa_r^\tau$ is a square matrix whose elements depend on the coefficients j_i and $H = diag \left(x_0^{-1} \right)$ is a positive diagonal matrix.

We shall denote by δ_k and Δ_k the leading principal minors of A and, respectively, $A + A^\tau$. By a leading principal minor of a matrix M we understand the determinant of a square submatrix that fits into the upper left-hand corner of M. Then we can prove the following propositions:

Proposition 1: A necessary condition for stability of the steady state of (5) is that $(-1)^k \delta_k > 0$ holds for all $k = \overline{1, m_0}$.

Proof. The proof is straightforward. Firstly, note that the system (5) is already in the reduced form, so all its eigenvalues are essential. We know that the steady state of (5) is stable if and only if all essential eigenvalues of AH are negative, that is AH is negative definite. This implies that A is a negative definite matrix, which further implies $(-1)^k \delta_k > 0$, $\forall k$.∎

This proposition will be helpful in determining when an extreme current is not stable in the sense of Definition 2, for any configuration of the kinetic parameters.

Proposition 2: A sufficient condition for stability of the steady state of (5) is that $(-1)^k \Delta_k > 0$ holds for all $k = \overline{1, m_0}$.

Proof. Remember that Δ_k are the principal leading minors of $A + A^\tau$. If $(-1)^k \Delta_k > 0$, $\forall k$, then we also have that $(-1)^k \Delta_k^{[h]} > 0$, $\forall k$, where $\Delta_k^{[h]}$ are the leading principal minors of $AH + HA^\tau$. Since $AH + HA^\tau$ is a symmetric matrix, the previous condition is equivalent to the fact that $AH + HA^\tau$ is negative definite matrix, which implies that AH is also a negative definite matrix. This means that all its eigenvalues are negative.∎

We also note that the condition $(-1)^k \Delta_k^{[h]} > 0$, $\forall k$ is equivalent to $A + A^\tau$ negative definite matrix.

If the network of interest is of a small order (usually less that 5), then the following theorem may be useful in studying the stability:

Proposition 3 (Hurwitz): A necessary and sufficient condition for stability is that the leading principal minors of the matrix

$$\mathcal{M} = \begin{pmatrix} a_1 & a_3 & a_5 & a_7 & \dots \\ 1 & a_2 & a_4 & a_6 & \dots \\ 0 & a_1 & a_3 & a_5 & \dots \\ 0 & 1 & a_2 & a_4 & \dots \\ \dots & \dots & \dots & \dots & \dots \end{pmatrix}$$

are all positive.

(Here a_i are the coefficients of the characteristic polynomial of AH, $P(\lambda) = \lambda^n + a_1 \lambda^{n-1} + a_2 \lambda^{n-2} +$

$$\cdots + a_{n-1}\lambda + a_n)$$

We propose the following algorithm for the stability analysis of a general biochemical network. The algorithm may give conditions among the rate parameters k_i's for which the whole network becomes stable or loses stability.

(1) Find the extreme currents, E_i, $i = \overline{1, q}$;

(2) For each extreme current determine the number of essential eigenvalues, using

no. essential eigenvalues = no. metabolites in the e.c. $- \dim(Ker(N_{r_i}^{\tau}))$.

Determine the reduced stoichiometric matrix, N_i, and the reduced kinetic matrix, κ_i, for each extreme current in part. Then write

$$A_i = N_i \operatorname{diag}(E_i)\,\kappa_i^{\tau}, \quad i = \overline{1, q}.$$

(3) Analize stability of each extreme current by looking at the eigenvalues of A_i. If one of the extreme currents, say E_α, is not stable in the sens of Definition 2, then it may induce the same feature in the whole network if the parameter j_α is large enough.

(4) We can check under which conditions that is possible, by looking at the principal leading minors of $A = N_r \cdot diag(\sum_{i=1}^{q} j_i E_i) \cdot \kappa_r^{\tau}$ and $A + A^{\tau}$;

(5) Find relations among j_i's for which all $(-1)^k \Delta_k > 0$, which are sufficient conditions for the stability of the network. These relations can be then transformed in relations among the k_i's.

(6) If there exists a k_0 such that $(-1)^{k_0} \delta_{k_0} < 0$, then the network may be unstable or present oscillations.

Examples

Example 1: (enzyme-catalysed reaction)

Let us consider the following enzyme-catalysed reaction, in which substrate is transformed by enzyme into product, via the formation of a enzyme-substrate complex:

$$\xrightarrow{k_i} S + E \underset{k_{-1}}{\overset{k_1}{\rightleftharpoons}} C \xrightarrow{k_2} E + P \xrightarrow{k_o}$$

where E, S, C and P are concentrations of, respectively, the free enzyme, the free substrate, the enzyme-substrate complex and the product.

Considering mass action kinetics, the time evolution of the system is governed by the following equations:

$$
\begin{aligned}
\frac{dS}{dt} &= k_i - k_1 S E + k_{-1} C, \\
\frac{dE}{dt} &= -k_1 S E + (k_{-1} + k_2)C, \\
\frac{dC}{dt} &= k_1 S E - (k_{-1} + k_2)C, \\
\frac{dP}{dt} &= k_2 C - k_o P
\end{aligned}
$$

This system satisfies a single conservation relation, $E + C = const.$, and has a unique steady state,

$$\hat{S} = \frac{k_1 k_M}{k_2 E_0 - k_i}, \quad \hat{C} = \frac{k_1}{k_2}, \quad \hat{P} = \frac{k_i}{k_o},$$

provided $k_2 E_0 > k_i$. Due to the conservation relation, one differential equation in the system (say, the second) is redundant. To determine the linearised stability of this steady state, then one must look at the eigenvalues of the reduced Jacobian. In the this case, the reduced Jacobian matrix is

$$J = \begin{pmatrix} -k_1 \hat{E} & k_{-1} & 0 \\ k_1 \hat{E} & -k_{-1} - k_2 & 0 \\ 0 & k_2 & -k_o \end{pmatrix}.$$

One can easily compute these eigenvalues,

$$-k_o, \quad \frac{-k_1}{2}\left[E_0 + k_M - \frac{k_i}{k_2} \pm \sqrt{(E_0 + k_M - \frac{k_i}{k_2})^2 - 4\frac{k_2 E_0 - k_i}{k_1}}\right],$$

and conclude that the system is stable, according to the above definition.

We now wish check stability using our method. The reduced stoichiometric matrix and the reduced kinetic matrix for the above reaction network are, respectively,

$$N_r = \begin{pmatrix} 1 & -1 & 1 & 0 & 0 \\ 0 & 1 & -1 & -1 & 0 \\ 0 & 0 & 0 & 1 & -1 \end{pmatrix}; \qquad \kappa_r = \begin{pmatrix} 0 & 1 & 0 & 0 & 0 \\ 0 & 0 & 1 & 1 & 0 \\ 0 & 0 & 0 & 0 & 1 \end{pmatrix}.$$

We have two stable extreme currents, $E_1 = [1, 1, 0, 1, 1]$ and $E_2 = [0, 1, 1, 0, 0]$. Their reduced Jacobians are:

$$A_1 = \begin{pmatrix} -1 & 0 & 0 \\ 1 & -1 & 0 \\ 0 & 1 & -1 \end{pmatrix}, \qquad\qquad A_2 = (-1)$$

from which we can check that, indeed, all eigenvalues are negative.
The matrix A is given by

$$A = N_r \cdot \mathrm{diag}(j_1 E_1 + j_2 E_2) \cdot \kappa_r^\tau = \begin{pmatrix} -j_1 - j_2 & j_2 & 0 \\ j_1 + j_2 & -j_1 - j_2 & 0 \\ 0 & j_1 & -j_1 \end{pmatrix}$$

and

$$H = \begin{pmatrix} \frac{k_2 E_0 - k_i}{k_i k_M} & 0 & 0 \\ 0 & \frac{k_2}{k_i} & 0 \\ 0 & 0 & \frac{k_o}{k_i} \end{pmatrix}$$

The necessary and the sufficient conditions in Propositions 1 and 2 are, respectively:

$$\delta_1 = j_1 + j_2 > 0, \quad \delta_2 = j_1(j_1 + j_2) > 0, \quad -\delta_3 = j_1^2(j_1 + j_2) > 0.$$

and

$$-\Delta_1 = 2(j_1 + j_2) > 0, \quad \Delta_2 = j_1(3j_1 + 4j_2) > 0, \quad -\Delta_3 = 2j_1^2(2j_1 + 3j_2) > 0.$$

All of them are satisfied for any configuration of the kinetic parameters, so that the network is stable.

Because the network is small, one can also use the Hurwitz theorem to analyse stability. In this case, the characteristic polynomial is

$$P(\lambda) = \lambda^3 + [(h_1+h_2+h_3)j_1 + (h_1+h_2)j_2]\lambda^3 + j_1(j_1+j_2)(h_1h_2+h_1h_3+h_2h_3)\lambda + j_1^2(j_1+j_2)h_1h_2h_3,$$

and one can easily check that the conditions for stability in Proposition 3 are satisfied.

Example 2 (*Oregonator*): We consider the Belousov-Zhabotynsky set of reactions:

$$A + Y \longrightarrow X + P$$

$$X + Y \longrightarrow 2\,P$$

$$A + X \longrightarrow 2\,X + 2\,Z$$

$$2\,X \longrightarrow A + P$$

$$B + Z \longrightarrow \tfrac{1}{2}f\,Y$$

Here f is a positive constant (the stoichiometric factor). There are no conservation relations in the system, so that

$$N_r = \begin{pmatrix} 1 & -1 & 1 & -2 & 0 \\ -1 & -1 & 0 & 0 & f/2 \\ 0 & 0 & 0 & 1 & -1 \end{pmatrix}, \qquad \kappa_r = \begin{pmatrix} 0 & 1 & 1 & 2 & 0 \\ 1 & 1 & 0 & 0 & 0 \\ 0 & 0 & 0 & 0 & 1 \end{pmatrix}$$

There are two extreme currents, $E_1 = [0, 1, 1, 0, 2]$ and $E_2 = [1, 0, 1, 1, 2]$. E_2 is stable and E_1 is not. The matrix A is given by:

$$A = N_r \cdot \mathrm{diag}(j_1 E_1 + j_2 E_2) \cdot \kappa_r^\tau = \begin{pmatrix} -3j_2 & -j_1+j_2 & 0 \\ -j_1 & -j_1-j_2 & f(j_1+j_2) \\ 2(j_1+j_2) & 0 & -2(j_1+j_2) \end{pmatrix}$$

The necessary conditions for asymptotic stability become:

$$-\delta_1 = 3j_2 > 0, \qquad \delta_2 = 3j_2^2+4j_1j_2-j_1^2 > 0, \qquad -\delta_3 = 2(j_1+j_2)[(f-1)j_1^2+3j_1j_2+(3-f)j_2^2] > 0 \quad (6)$$

and, the sufficient conditions are:

$$-\Delta_1 = 6j_2 > 0, \qquad \Delta_2 = -4j_1^2 + 16j_1j_2 + 11j_2^2 > 0 \quad \text{and} \tag{7}$$

$$-\Delta_3 = 2(j_1 + j_2)[4(f-3)j_1^2 + (24+2f-3f^2)j_1j_2 + (18-2f-3f^2)j_2^2] > 0.$$

We can observe that, from (6) and (7) one can find conditions upon f and the j_i's such that the steady state is stable or not. We shall discuss here in more detail only the case $f = 1$, when the system has a unique steady state. In this particular case, the matrix A becomes

$$A = \begin{pmatrix} -3j_2 & -j_1+j_2 & 0 \\ -j_1 & -j_1-j_2 & j_1+j_2 \\ 2(j_1+j_2) & 0 & -2(j_1+j_2) \end{pmatrix}$$

The above necessary and sufficient conditions become, respectively,

$$-\delta_1 = 3j_2 > 0, \qquad \delta_2 = 3j_2^2 + 4j_1j_2 - j_1^2 > 0, \qquad -\delta_3 = 4j_2[2j_1^2 + 3j_1j_2 + j_2^2] > 0$$

and

$$-\Delta_1 = 6j_2 > 0, \qquad \Delta_2 = -4j_1^2 + 16j_1j_2 + 11j_2^2 > 0, \qquad -\Delta_3 = -16j_1^3 + 30j_1^2j2 + 72j_1j_2^2 + 26j_2^3 > 0.$$

Since $v = \sum_i j_i E_i$, we cand write the reaction rates in terms of j_i's:

$$\begin{aligned} v &= \left(k_1 aY, \, k_2 XY, \, k_3 aX, \, k_2 X^2, \, k_5 bZ\right)^\tau & (8) \\ &= \left(j_2, \, j_1, \, j_1 + j_2, \, j_2, \, 2j_1 + 2j_2\right)^\tau & (9) \end{aligned}$$

We now change the basis in the steady state space from $\left\{k_1 aY, \, k_2 XY, \, k_3 aX, \, k_2 X^2, \, k_5 bZ\right\}$ to a Gröbner basis. We can do this using *Matematica*, and we obtain:

$$Gr\ddot{o}bner \; basis: \quad \{k_3^2 v_4 - k_4 v_3^2, \, k_2 k_3 v_1 v_4 - k_1 k_4 v_2 v_3, \, k_1 k_3 v_3 - k_2 v_1 v_3, \, k_2^2 v_1^2 v_4 - k_1^2 k_4 v_2^2\}$$

The steady states have to verify the conditions:

$$k_3^2 v_4 - k_4 v_3^2 = 0; \quad k_2 k_3 v_1 v_4 - k_1 k_4 v_2 v_3 = 0; \quad k_1 k_3 v_3 - k_2 v_1 v_3 = 0; \quad k_2^2 v_1^2 v_4 - k_1^2 k_4 v_2^2 = 0,$$

from which we get:

$$\begin{cases} j_1 = k_2 j, & (j - \text{ positive parameter}) \\ j_2 = k_1 k_4^{1/3} j^{2/3} \end{cases}$$

We can easily see that, if j_2 is large enough (i.e. k_2 is large enough), then $\Delta_2, \, -\Delta_3 > 0$ and this implies stability in the sense of Definition 2.

On the other side, if j_1 is large enough (i.e. either k_1 or k_4 is large enough), then $\delta_2 < 0$ which implies the existence of an eigenvalue with $\mathcal{R}e(\lambda) \geq 0$, thus the system may present oscillations or instabilities

Conclusions

The stability of steady states in a biochemical network is a very important issue. In this paper, we have presented some conditions for the stability or instability of the steady state in a biochemical network, when that steady state exists. In the approach, we have used the method of splitting up the whole netwok into subnetworks generated by elements of a basis in the steady state space, called the extreme currents. As shown in the examples presented, from these conditions one may obtain relations among the kinetic parameters for which stability or instability occur.

Acknowledgements

I would like to thank the UniNet and the Klaus Tschira Foundation for financial support, and to Ursula Kummer and Tim Johann for many valuable discussions we have had.

Bibliography

[1] B. Clarke, Stability of complex reaction networks, in I. Prigogine and S. Rice (Eds.), Advances in Chemical Physics, New York Wiley Vol. 43, pp 1–216, 1980.

[2] B. Clarke, Stoichiometric network analysis, Cell Biophysics, Vol. 12, pp 237–253, 1988.

[3] Karin Gatermann, Markus Eiswirth, Anke Sensse: Toric ideals and graph theory to analyze Hopf bifurcations in mass action systems. J. Symb. Comput. 40(6): 1361–1382, 2005.

[4] A. Sensse, Convex and toric geometry to analyze complex dynamics in chemical reaction systems, PhD thesis, Otto-von-Guericke-Universität Magdeburg, 2005.

Neuronal Networks

Gamma oscillations in interneuron networks: a combined experimental-computational approach

Imre Vida[1], and Marlene Bartos[2]
[1]Institute of Anatomy and Cell Biology and
[2]Institute of Physiology, University of Freiburg,Germany

Abstract

Oscillations provide temporal structure for information processing in neuronal networks. In particular, oscillations in the gamma frequency band (30 - 90 Hz) are thought to be important for higher brain functions, such as sensory binding and attention. Generation of these oscillations depends critically on mutual synaptic interactions among fast-spiking soma-inhibiting interneurons, so called basket cells. Previous experimental and theoretical studies suggested that properties of basket cell-basket cell synapses define coherence and frequency of gamma activity. It was proposed that slow, weak and hyperpolarizing synapses promote gamma oscillations in interneuron networks. However, gamma activity in these networks is very sensitive to changes in the synaptic parameters and to heterogeneities. Performing whole-cell patch-clamp recordings, we have recently found that the kinetics at basket cell-basket cell synapses is unexpectedly fast and have shunting, rather than hyperpolarizing effect. Computational analysis further showed that a network model based on experimentally defined properties generates highly coherent gamma activity over a large area of the parameter space with increased tolerance against heterogeneities. Thus, results of this combined experimental-computational study suggest that specialized synaptic properties turn cortical basket cell networks into robust gamma frequency oscillators.

Introduction

Gamma activity (30 - 90 Hz) is an oscillatory pattern observed throughout the brain. These oscillations are thought to serve as reference signals for temporal coding of information in neuronal networks [3, 9]. Gamma oscillations have been examined extensively in the hippocampus where they occur typically in combination with theta frequency oscillations during explorative behavior [4]. Previous experimental evidence indicates that GABAergic inhibitory interneurons play a key role in the generation of these oscillations. Interneurons, in particular fast-spiking, soma-inhibiting cells (basket cells, BC), fire action potentials at high-frequency, phase-locked to the gamma waves in the hippocampus both *in vivo* [4] and *in vitro* [8, 10]. Furthermore, gamma oscillations generated in hippocampal slices *in vitro* by sustained metabotropic receptor activation are completely blocked by $GABA_A$ receptor antagonists [6, 19].

Consistent with this hypothesis, previous computational studies showed that interneurons can synchronize at various frequencies through mutual inhibitory synapses [14, 16, 18] and entrain the activity of principal cell population *via* their divergent inhibitory output synapses [11]. In interneuron network models with slow, weak and hyperpolarizing synapses, gamma oscillations were reliably generated in response to a tonic excitatory drive [14, 18]. However, oscillations in these networks were very sensitive to changes in the synaptic parameters [18]. Additionally, oscillations emerged only when heterogeneity of the excitatory drive was low (<5 %; [18] see also [13]). The high sensitivity of the models to heterogeneities in the drive was inconsistent with the experimentally-observed large heterogeneity of metabotropic responses in interneurons (30%; [15]).

In summary, results of prior experimental and computational studies underscored the role of mutual inhibitory synaptic connections among interneurons in the generation of gamma activity. However, the proposed models could not explain the robustness of oscillations observed *in vivo* and *in vitro*. Furthermore, it was unclear whether assumptions of the models were compatible with properties of interneuron-interneuron synapses in cortical circuits.

Physiological characterization of basket cell-basket cell synapses

In order to determine functional parameters of BC-BC synapses, we have performed whole-cell patch-clamp recordings from synaptically coupled pairs of fast-spiking BCs in acute hippocampal slices at near physiological temperature [1, 2]. The results showed that inhibitory postsynaptic currents (IPSC) mediated by $GABA_A$ receptors at morphologically identified BC-BC synapses have unexpectedly fast kinetics and high amplitude (Fig. 1A). The unitary IPSCs recorded in the voltage-clamped postsynaptic BCs had short synaptic delay, rapid rise and decay. The decay time constant was in the range of 1.2 - 2.5 ms, significantly shorter than what was assumed earlier (∼10 ms). Furthermore, IPSCs had large amplitude and followed presynaptic action potentials with high reliability. Similar results were obtained for BC-BC synapses in the neocortex [7], suggesting that precisely timed fast signaling and strong coupling is a general principal in BC networks throughout the cortex.

Next, we investigated whether inhibition had a hyperpolarizing effect in BCs [17]. To avoid perturbation of the intracellular chloride concentration, perforated-patch-clamp recordings were obtained from the interneurons. Pharmacologically isolated IPSCs were evoked by extracellular stimulation of presynaptic axons in the cell body layer at various holding potentials in order to determine the membrane potential at which the polarity of the inhibitory responses reverses (reversal potential). The reversal potential of the $GABA_A$ receptor-mediated IPSCs in BCs had a mean value of -52 mV, between resting membrane potential and action potential threshold, indicating that inhibition had a shunting, rather than hyperpolarizing effect in these neurons [17].

In conclusion, the experimental data showed that BC-BC synapses have highly specialized properties, very different from those assumed in previous interneuron network models.

Computational analysis of interneuron networks with experimentally determined synaptic properties

Do fast, strong and shunting synapses support the generation of gamma oscillations in interneuron networks? In previous models, coherent gamma oscillations were generated with weak, slow

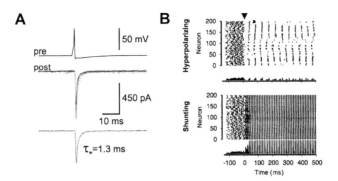

Figure 1: Mutual inhibition among hippocampal BCs promote gamma oscillations. **A** Action potentials in a presynaptic BC (top) elicit IPSCs of high amplitude and fast kinetics in the voltage-clamped postsynaptic BC in the CA1 area (middle, 6 traces superimposed; bottom, average IPSC). **B** Raster plots of activity pattern in the interneuron network models with fast, strong and hyperpolarizing (top) or shunting synapses (bottom).

and hyperpolarizing synapses. If one of the synaptic parameters was changed, oscillations were abolished [18]. However, a combined change of the parameters may bring unexpected results. To test this possibility, we have created an interneuron network model based on the experimentally derived synaptic parameters [17]. The network, implemented in the *Neuron* simulation environment, consisted of 200 single compartment neurons with modified Hodgkin-Huxley type sodium and potassium conductances [18]. Synaptic connections were assigned randomly, using a spatial rule to mimic the anatomical connectivity pattern of BCs. To imitate sustained metabotropic receptor activation, a tonic excitatory drive with heterogeneous amplitude distribution was applied to the neurons.

Unexpectedly, the simulations showed that in the interneuron network model with realistic synaptic properties, oscillations were not abolished but enhanced compared to those in previous network models. Activation of fast and strong inhibitory synapses during incoherent activity resulted in a rapid synchronization and a sustained oscillatory activity in the network (Fig. 1B)[1, 2]. This effect was independent of the reversal potential of inhibition, but was critically dependent on synaptic delays [2, 5, 12]. Analysis of the network with shunting inhibition, further revealed an enhanced gamma activity at low excitatory drive (1B, bottom) and an improved robustness of the oscillations against heterogeneities in the drive. These effects were due to a differential, excitation level-dependent influence of shunting inhibition on the neurons: weakly excited neurons are facilitated, whereas strongly excited neurons are inhibited. This differential influence can directly explain the enhanced network activity at low excitation levels. Furthermore, in a network exposed to a heterogeneous excitatory drive, shunting inhibition will homogenize the firing frequencies of the neurons and increase thereby the robustness of oscillations [17].

Conclusions

We have addressed the role of mutual inhibitory synapses among GABAergic interneurons in the generation of gamma oscillations by a combined electrophysiological, neuroanatomical and computational approach. The electrophysiological data revealed that BC-BC synapses are fast, strong and shunting, rather than slow, weak and hyperpolarizing as previously assumed. Computational analysis further showed that the interneuron network model with realistic synaptic properties generate coherent gamma oscillations with increased robustness against heterogeneities. Thus, specialized properties of inhibitory synapses in interneuron networks may underlie robust gamma-frequency oscillations in cortical circuits.

Bibliography

[1] Bartos & al., *J Neurosci*, 21:2687-98, 2001.

[2] Bartos & al., *Proc Natl Acad Sci U S A*, 99:13222-7, 2002.

[3] Buzsáki & Draguhn. *Science*, 304(5679):1926-9, 2004.

[4] Bragin & al., *J Neurosci*, 15:47-60, 1995.

[5] Brunel & Wang, *J Neurophysiol*, 90:415-30, 2003.

[6] Fisahn & al., *Nature*, 394:186-9, 1998.

[7] Galarreta & Hestrin, *Proc Natl Acad Sci U S A*, 99:12438-43, 2002.

[8] Gloveli & al., *J Physiol*, 562:131-47, 2005.

[9] Gray & Singer, *Proc Natl Acad Sci U S A*, 86:1698-702, 1989.

[10] Hájos & al., *J Neurosci*, 24:9127-37, 2004.

[11] Lytton & Sejnowski, *J Neurophysiol*, 66:1059-79, 1991.

[12] Maex & De Schutter, *J Neurosci*, 23:10503-14, 2003.

[13] Tiesinga & Jose, *Network*, 11:1-23, 2000.

[14] Traub & al., *J Physiol*, 493:471-84, 1996.

[15] van Hooft & al., *J Neurosci*, 20:3544-51, 2000.

[16] Van Vreeswijk & al., *J Comput Neurosci*, 1:313-21, 1994.

[17] Vida & al., *Neuron*, 49:107-17, 2006.

[18] Wang & Buzsáki, *J Neurosci*, 16:6402-13, 1996.

[19] Whittington & al., *Nature*, 373:612-5, 1995.

Networks of neurons

Demian Battaglia
Laboratory of Neurophysics and Physiology
CNRS UMR 8119 , University René Descartes ,
45, Rue des Saints Pères - 75270 Paris cedex 06, France.

Abstract
In this short chapter, we will try to review very briefly some of the models that have been proposed in the literature to describe the different key components of neural systems. In the title, "Networks of neurons", we deliberately inverted the typical order of the words. This is meant to be a prelude to the "graph-theory-inspired" organization of the contents that we will try to follow in the next pages (nodes, links and then the whole graph), in the attempt to highlight the affinities existing with the other network theories presented in this same scientific report.

According to the greek physician Galen, the brain was thought to be a cold and moist gland, secreting fluids conveyed then to the periphery of the body by the nerve fibers. Only the microscope revealed the true nature of brain tissue. The first detailed observations of nerve cells belong to the late nineteenth century and are exposed in the pioneering works by Camillo Golgi and Santiago Ramón y Cajal. Using the Golgi staining, the spanish histologist could clearly identify separated cell bodies connected by prominent branching projections. His studies contributed then to establish the description of the nervous tissue as a complex network of interconnected and distinct signaling units, that received since then the name of neurons.

Nodes: neurons

At every site of the "neural graph" sit highly specialized cells that come in many different varieties, differing widely in appearance and electrochemical properties. Neurons are typically 4 to 100 μm in diameter. The nucleus is located in the central part of the cell, the soma, but the most distinctive feature of neuronal cells are the extended processes which they use to send and receive information. Most neurons have input dendritic trees with profuse dendritic branches. In addition, a single output axon is used to send electrical signals to many target cells, touched by its multiple ramifications. A large number of different kinds of passive and active channels allows ions (mainly Na^+, K^+, Ca^{2+} and Cl^-) to move across the membrane (otherwise insulating and acting as a capacitor). Setting conventionally the extracellular potential to be at 0 mV, most of the time an excess of negative charges will cause the internal potential to be negative (at approximately -70 mV), as a result of the net equilibrium between the outward and inward fluxes of ions. Current flowing into the cell increases the resting membrane potential, a process called depolarization.

When the membrane potential is raised above a certain threshold level, a positive feedback process is initiated and a large voltage fluctuation up to +40 mV is suddenly generated and propagated even at very large distance.

Models that describe the membrane potential just in terms of a single variable V are called single-compartment models [8, 19, 20]. Multi-compartment models, in which a single neuron itself is split into a network of sub-neuronal compartments (for instance different sections of a large dendritic tree), could be introduced as well, allowing to analyze spatial variations of the membrane potential and then the detailed effects of an intricate morphology [20, 19]. In both cases, the electrical properties of a neuronal model will depend on the precise kinds of nonlinear voltage-dependent conductances that will be present [14, 9]. Channel gating mechanisms involve complex conformational changes of the proteins constituting the channels themselves. Single channel current recordings have shown that a channel switches very quickly and in a stochastic-like manner between the open and the closed state [30]. Probabilistic Markov-chain models of individual channels have been proposed in order to describe their switching dynamics and the results have been found to be in agreement with previously introduced empiric laws providing a purely phenomenological expression for the fraction of open channels[28, 23]. In the 1950s Hodgkin and Huxley, studying the generation of action potentials in the giant squid axon [15], proposed a first description in terms of nonlinear differential equations. The membrane current was just obtained as the sum of a leakage current, a delayed-rectified K^+ current and a transient fast Na^+ current. A four-dimensional system of ODEs was needed In order to describe the complex nonlinear dynamics of the membrane potential and of the gating variables governing the opening of the channel conductances for a single neuron. But larger dimensions can easily be achieved. A whole zoo of different ion channels have been identified and various phenomena like postinhibitory rebound and bursting, adaptation, etc. can be taken into account[8]. Furthermore, in multi-compartment models, each compartment will have a separate membrane potential variable, governed by an equation similar to the original Hodgkin-Huxley model. In this way, it is not difficult to believe that single cells *conductance-based models* using thousands of differential equations can easily be constructed.

Such models can quickly become unmanageable when going beyond the study of single cell properties Powerful methods like phase-plane analysis can be used to study the dynamical oscillatory properties of two-dimensional differential equation systems. Systematic techniques for the elimination of two of the four original variables of the Hodgkin-Huxley equation have therefore been developed along the years, mainly based on the observation that different time-scales act simultaneously [12, 18]. We will just cite here the FitzHugh-Nagumo model and the Morris-Lecar model (reviewed in [12]). A completely different class of simplified cell models can be obtained by considering that, basically, all the spikes are identical among them. It is then tempting to neglect completely the biophysical mechanisms responsible for the generation of action potentials. In the *Integrate-and-Fire* models a stereotypical action potential occurs whenever the membrane potential crosses a threshold value V_{th}. At this point a δ-like spike is simply "glued" to the previously integrated sub-threshold trajectory and the potential is artificially reset to a value V_{reset}. With these approximations, the resulting leaky integrate-and-firemodel neuron behaves just like an electric circuit with a capacitor and a leakage conductance in parallel. It is then described by the very simple RC-circuit differential equation augmented with the fire-and-reset rule. Integrate-and-fire models are easy to manage and allow for still considerably sophisticated analytical analysis. Furthermore, features of neuronal firing like refractoriness, firing-rate adaptation or other nonlinear effects can be easily incorporated without affecting the model inherent simplicity [8, 12, 20].

Nevertheless, for a systematic analytical and numerical investigation of the collective behavior of large systems, even more simplified caricatures of the original neuron can become extremely helpful. In the so-called *rate models* the state of a network unit is described just by a single positive continuous real variable which we will call the activity [34, 33, 1, 20]. The precise interpretation of this variable can differ according to the context (firing time-rate, population rates —*i.e.* number of neurons firing in a population—, instantaneous activity of slow synapses, etc.). In most cases, the evolution of the activity variables is given by very simple equations of the form: $\tau \frac{dm}{dt} = -m + \Phi[h_{in}]$, where h_{in} represent an input field (resulting from external stimulation current as well as from recurrent synaptic connections) and $\Phi(\cdot)$ is a nonlinear input-output transfer function of the unit (threshold-linear or sigmoidal are two common choices).

The all-or-none properties of neuronal firing may suggest the radical choice to model individual neuronal units as *binary* on-or-off elements. Time is divided in cells of the size of a refractory period, to exclude the generation of more than a single spike per cell during each step. The resulting Mc-Culloch and Pitts neuron resembles then a discrete-time dependent boolean or physical spin variable [1, 2, 10]. The original goal in 1943 was to show that logical control circuitries can be implemented by small networks of elementary neuron. Since then, binary neuron models have been widely used in the literature of artificial neural networks [2, 3, 10]. The physical analogy with disordered spin systems allowed to use systematically powerful statistical mechanics techniques [10, 1] to investigate large network properties like the number of attractor fixed-points, synchronization, etc. Until now, neuronal models simpler than the binary neuron have not been proposed in the literature and it is quite unlikely that they will ever appear in the future.

Edges: dendrites, axons and synapses

Neurons are shaped to create communication links. A dendritic tree made up of 10 to 400 different dendritic tips can carry up to 200000 synapses. A single axon can be long up to 1 m (in the case of motoneurons) and can innervate up to 10000 output cells with its ramifications [3]. Dendrites and axons acts primarily as transmission lines for electric signals, but several details of their structure can be connected to additional functions [19, 20, 25]. The axon is essentially a cable designed to propagate at long distance the action potentials generated at the axon hillock close to the cellular body. Myelinated axons act as an active repeating transmission line, regenerating the propagated signal at regular spatial intervals in order to prevent its attenuation. Propagation inside an axon is (usually) *directed*, a relevant detail from our graph-theoretical point of view. The propagation of post-synaptic potentials in the dendritic tree is most of the time passive but can be as well highly non trivial. Cable theory and useful techniques developed by Rall can be used to predict details of the postsynaptic potential propagation in a branched dendritic tree [8, 20]. It turns out that dendritic trees are electrically distributed rather than isopotential elements, and voltage spatial gradients are created as a consequence of locally applied synaptic inputs. The large attenuation in the periphery-to-body propagation ensures that the large depolarization needed for the generation of an axon potential can be build up only out of a large number of excitatory inputs, allowing for a greater *robustness*. Furthermore, the integration of the many afferent signals is affected by the morphology of the tree and by the precise spatio-temporal pattern of application of the inputs. Dendritic propagation can then actually perform computations by itself (detect direction of motions, act as cohincidence detectors, implement multidimensional classification tasks , etc., see [3, 25] for a summary)

Synapses can be seen on the other hand as the "plugs" joining together different "neural wires" and the coupling is realized in a complex electro-chemical way. When an action potential enter a presynaptic terminal it activates voltage-dependent Ca^{2+} channels. Several vesicles containing transmitter molecules fuse then with the cell membrane, allowing the released chemical messengers to bind receptors on the post-synaptic side of the synaptic cleft. Sometimes the effect of the formation the transmitter-receptor complex is direct and brings directly to the opening of ion channels (ionotropic receptors). Other times the opening of channels will be triggered only indirectly by the activation of an intracellular cascade of metabolic signaling (metabotropic receptors). In general different receptors will induce postsynaptic potentials of different strength and time-course [8, 9, 39]. The probability of transmitter release and the relative strength of the post-synaptic response can depend on the previous activity of the synapse. Several time scales come into play: short-term plasticity is relative to effects on the synaptic efficacy lasting from milliseconds to tenths of seconds [40, 37], while long-term plasticity can result in changes lasting for a whole life [21, 4, 22, 26]. Such modifications are usually considered to be in relation with the formation of memories. Plasticity can be modeled by making the maximum synaptic efficacy a function of time [8, 20]. First-order kynetics equations are often used to describe short-term effects. More recently, experimental studies have pointed out that facilitation and synaptic depression happen at the level of individual synapses in a quantized way, involving just switchings between a small number of discrete internal states with different efficacies [29, 11].

Network topology

The human brain contains between 7×10^{10} and 8×10^{10} neurons, most of them belonging to to the cerebellar (5×10^{10}) and the cerebral cortex ($\sim 1.5 \times 10^{10}$). The main fiber bundle connecting the two hemispheres, the corpus callosum, contains approximately 10^8 fibers [3]. Even focusing to the neocortical tissue only, the extraction of the adjacency matrix of such a large and complex graph seems definitely to be an impossible (and, eventually, useless) task. The scenario is complicated by the fact that most connections are not one-to-one, many neurons receiving input from thousands of other ones and transmitting their output to as many. Crude maps of the cortical connectivity can nevertheless be obtained by anatomical and finer statistical observations.

One general principle which appear to be respected is the separation of the cortical tissue in zones performing the actual computations and others whose main role is just to maintain very-long-range connections between different part of the brain. Apart from blood vessels and supporting cells, the *white matter* is indeed made up only of myelinated axons. The *gray matter* contains on the other hand all the dendritic trees (and therefore the synapses), shorter unmyelinated axons and cell bodies. The distinction between white and gray matter is macroscopically visible (appearing actually in brain sections as regions of different color). Most of the white matter belongs to the cortex, and only a small fraction projects toward other parts of the brain. We might actually say that "one of the basic principles of brain connectivity is to tightly connect the cortex with itself" [3]. Several authors have indeed proposed arguments to show that the brain volume would become at least a 50% larger if we had to mantain the same degree of cortical connectedness in absence of the segregation between white and gray matter. It is possible to find in the literature various other examples in which the precise wiring among areas or subareas is shown to optimize some suitable cost function (energy consumption, transmission delay, etc., see [3, 7]).

A second general architectonical observation is that the cortex is an essentially bidimensional 2 mm-thick system. It is traditionally divided into six parallel layers, numbered progressively from the surface to the inner side, lying on the white matter. Considering the statistically more represented patterns of inter-layer connections, a typical directed path would enter the cortex from layer IV, which receives the sensory inputs, then it would go up to III, down to V or VI, and then back out the cortex to some sub-cortical structure [38, 5, 36]. Neurons which respond to similar stimuli (*e.g.* flashing bars at a given location and of a given orientation) tend to be vertically arranged in the cortex, forming 200–300 μm wide cylinders known as *cortical columns* and constituting a widely repeated modular structure [27]. Barrels in somatosensory cortex and ocular dominance columns in visual cortex have fairly discrete boundaries with neighboring columns . In other cases, like for instance orientation columns, there is a smooth variation in response properties moving parallel to the cortical surface. Columns with different orientation preferences, for instance, are spatially organized around singular points, called pinwheels, that are adjacent to columns of every possible orientation. The whole set of this neighboring columns responding to visual stimuli of all the possible orientations (in a precise receptive field) can itself be considered as a functional module, usually referred to as a *hypercolumn* [24, 27]. The cortical columns can also be identified by looking at the variation of the density of connections in anatomical sections. It is possible indeed to identify spatially confined vertical densities of connections that may provide the base for purely intra-columnar —*i.e.* local— information processing. In addition, there are long-range (generally up to a few millimetres long), sparse and patchy connections, which tend to connect among them spatially separated columns with a same feature preference, like the same orientation or ocular dominance preference [6, 13, 27].

Unfortunately a reconstruction of the microcircuitry at the level of a single column is still beyond the technical possibilities of observation, and statistical methods have to be used in order to unravel the details of local connectivity [5, 36]. A further complication is given by the fact that local connections could be actually *evolving in time*, because of dendritic and synaptic growth. There are evidences that is impossible for an axon of a pyramidal neuron to avoid to establish at least a *geometric* contact (defined just in terms of observed spatial touch-points) with any other pyramidal neuron in a local neighborood of 100-200 μm from the soma. It is not on the other hand needed that every geometric contact-point correspond to an effective synaptic coupling, able to trigger post-synaptic responses when the presynaptic cell is active [17]. Several observations suggest that the effective connectivity is in reality rather sparse (only 15% of the potential synaptic sites appeared to be effective in recent studies over cortical slices), while the geometric connectivity appears to be essentially all-to-all. This could allow for low-energy-consuming local rewirings, since new effective circuits could be activated just by growing or retracting few synaptic spines (*tabula-rasa hypothesis*, [17]). Statistical sampling of the effective connections can also bring to the observation of *local motifs* consistently repeated across different cortical slides [35]. Several connectivity patterns involving bidirectional connections among two or three neurons are over-represented with respect to what would be expected if the local network was completely random. Furthermore, stronger efficacy connections are found to belong to such clustering motifs with a larger probability than weaker connection [35]. An interesting open question is to establish whether or not such features of the local connectivity are relevant from a dynamical and information processing view-point.

Conclusion

In this review chapter we analyzed several ways of modeling the nodes and the links of neural networks in a descending order of complexity. We finally had a fast look to what is known about the overall geometry of the connectivity of the brain, in terms of general organizational principles and more precise statistical observations about the micro-anatomy of the tissue. The presented descriptions of the neural components ranged from highly simplified toy-models to detailed characterizations involving large systems of non-linear differential equations. Choosing the right model requires always a careful evaluation of the desired goals. In particular, if it is true that excessively simple models could bring to unbiological results, it is true as well that the excess of inessential details could distract from the investigation of the most relevant processes. To conclude, the modeler's aim should be to always identify the simplest model capturing the complex behaviors we are interested in with the minimal effort.

Bibliography

[1] Amit D., *Modeling brain function*, Cambridge University Press (1992)

[2] Anderson J.A., Rosenfeld E., *Neurocomputing: foundations of research*, MIT Press, Cambridge (MA) (1988)

[3] Arbib M.A. *editor, The handbook of brain theory and neural networks*, MIT Press, Cambridge (MA) (1995)

[4] Bi G., Poo M., *Ann. Rev. Neurosci.* 24, 139–166 (2001)

[5] Braitenberg V., Schuz A., *Anatomy of the Cortex: Statistics and Geometry*, Springer-Verlag, Berlin (1991)

[6] Bressloff P.C., Cowan J.D., *J. Physiol. Paris* 97:221-36 (2003)

[7] Chklovskii D., Schikorski T., Stevens C.F., *Neuron* 34:341–347 (2002)

[8] Dayan P., Abbott L.F., *Theoretical neuroscience, computational and mathematical modeling of neural systems*, MIT press, Cambridge (MA) (2001)

[9] Destexhe A., Mainen Z., Sejnowsky T., *J. Comput. Neurosci.*, 1, 195–230 (1994)

[10] Engel A., Van der Broeck C., *Statistical mechanics of learning*, Cambridge University Press (2001)

[11] Fusi S., Drew P.J., Abbott F.L., *Neuron* 45:599-611 (2005)

[12] Gerstner W., Kistler W., *Spiking neuron models*, Cambidge University Press (2002)

[13] González-Burgos G., Barrionuevo G., Lewis D.A., *Cerebral Cortex* 10:82 (2000)

[14] Hille B., *Ionic channels of excitable membranes*, Sinauer associates, Sunderland (1992)

[15] Hodgkin A.L., Huxley A.F., *J. of Physiol.* 117:500–544 (1952)

[16] Hopfield J.J., *Proc. Nat. Acad. Sci.* 79:2554 - 2588 (1982)

[17] Kalisman N., Silberberg G., Markram H., *Proc. Nat. Acad. Sci.* 102:880-885 (2005)

[18] Kepler T.B., Abbott L.F., Marder E., *Biol. Cybern.* 66:381–387 (1992)

[19] Koch C., *Biophysics of computation: Information processing of single neurons*, Oxford University Press (1998)

[20] Koch C., Segev I. *editors, Methods in neuronal modeling: from synapses to networks*, MIT press, Cambridge (MA) (1998)

[21] Lynch M.A., *Physiol. Rev.*, 84: 87136 (2004)

[22] Markram H., Lubke J., Frotscher M., Sakmann B.,*Science*, 275:213–215 (1997)

[23] Marom S., Abbott L.F., *Biophys. J.* 67:515–520 (1994)

[24] Martin K.A.C., *Q. J. Exp. Physiol.* 73:637 (1988)

[25] McKenna T., Davis J., Zornetzer S.F. *editors, Single neuron computation*, Academic Press, Boston (1992)

[26] Paul Miller P., Zhabotinsky A.M., Lisman J.E., Wang X.-J., *PLOS biol.* 3 (2005)

[27] Mountcastle V.B., *Brain* 120:701 (1997)

[28] Patlak J., *Physiol. rev.* 71:1047–1080 (1991)

[29] Petersen C.C.H., Malenka R.C., Nicoll R.A., Hopfield J.J., *Proc. Natl. Acad. Sci* 95:4732–4737 (1998)

[30] Sakmann B., Neher E., *Single channel recording*, Plenum, New York (1983)

[31] Senn W., Fusi S., *Neur. Comput.* 17: 2106–2138 (2005)

[32] Shapley R., Hawken M., Ringach D.L., *Neuron* 38:689-99 (2003)

[33] Shriki O., Hansel D., Sompolinsky H., *Neural Comput.* 15:1809–1841 (2003)

[34] Sompolinsky H., White O.L., Theory of Large Recurrent Networks: From Spikes to Behavior, in *Methods And Models In Neurophysics: Proceedings of the Les Houches Summer School 2003*, Elsevier (2004)

[35] Song S., Sjöström P.J., Reigl M., Nelson S., Chklovskii D.B., *PLOS Biol.* 3 (2005)

[36] Thomson A.M., West D.C., Wang Y., Bannister A.P., *Cereb. Cortex* 12:936–953 (2002)

[37] Tsodyks M.V., Markram H., *Proc. Nat. Acad. Sci.* 94:719–723 (1997)

[38] White E.L., *Cortical circuits*, Birkhauser, Boston (1989)

[39] Zimmermann H., *Synaptic Transmission: Cellular and Molecular Basis*, Oxford University Press (1994)

[40] Zucker R.S., *Ann. Rev. Neurosci.* 12:13–31 (1989)

Chaos in networks with delayed local inhibition

Demian Battaglia and David Hansel
Laboratory of Neurophysics and Physiology,
CNRS UMR 8119 , University René Descartes ,
45, Rue des Saints Pères - 75270 Paris cedex 06, France.

Abstract

It has been shown recently that the introduction of explicit delays generates a very rich repertoire of different dynamical states (uniform oscillations, standing waves, chaos, etc.) in large networks which are analogue to the classical "ring model" for feature selectivity in local cortical circuits. It can be shown that the complexity of the phase diagram is substantially preserved when the size of the system is reduced to a very small number of neurons and we will focus in particular on the mechanisms underlying the transition to chaos in the smallest possible ($N = 2$) ring model. We show that chaotic oscillations are always found in presence of strong delayed self-inhibition if the strength of the bidirectional excitatory connections exceeds a small critical value. This chaotic phase is very robust and does not depend on the delay of the reciprocal interaction. The onset of chaos (via a Feigenbaum period-doubling scenario) is preceded by a characteristic transition (at the point of perfect decoupling among the two units) toward a periodic phase exhibiting phase-locking at an intermediate phase-shift between 0 and π. This spontaneous symmetry breaking allows the leading neuron to act as an oscillatory forcing driving to chaos the laggard unit.

Report

There are both electrophysiological and anatomical evidences of the existence of a spatial organisation of the cortex in functional modules (see introductory chapter). An example already mentioned is represented by primary visual cortex (V1) where cells exhibit selective responses to elongated stimuli with different orientations and where cells with similar preferred orientation tend to interact more among them than with those who respond to different orientations. Modelling studies have shown that spatially modulated excitation and inhibition brings to complex spatio-temporal patterns of activation [1, 2, 3], which could be associated to functions like feature selectivity [4, 2, 5] and spatial working memory [6, 7].

In order to gain insight on the dynamics of large cortical networks, firing-rate models constitute a very effective tool [4, 2, 8, 9]. But many of the previous analyses have neglected an important property of neural interaction, that is the existence of delays, on the order of milliseconds, due to the finite velocity of axonal propagation, to dendritic and synaptic processing [10], as well to the

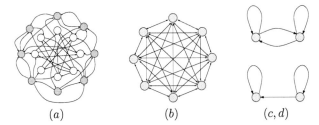

$$(a) \qquad\qquad (b) \qquad\qquad (c,d)$$

Figure 1: Graph topologies of the conductance-based model (a) and of the associated rate model (b); the simplest possible $N = 2$ ring-model (with feedback (c) and with excised feedback (d)).

intrinsic non-linearity of spike-generation [11]. It is known that delays can induce homogeneous oscillations in inhibitory networks with homogeneous random connectivity [12, 13]. The combined effect of delays and spatially structured interactions has been studied only recently [14, 15], and it has been seen to induce a large repertoire of possible dynamical states. In particular, chaotic oscillations have been observed in presence of strong inhibitory short-range connections and diluted long-range excitation, differently from previous results (chaos had been until now reported for mexican-hat connectivity [16] or in the case of balanced networks [17, 18]). The modeling results could then be relevant in order to explain the destabilization of regular rhythms arising in inhibitory interneuron networks [19, 20]. Let us study these proposed models in more detail.

Two distinct populations of conductance-based neurons (of the Wang-Buzsaki type [20]) are first considered, one excitatory and one inhibitory. Synaptic delays are explicitly introduced. The network is 1-D with periodic boundary conditions and each neuron in each one of the populations can be labeled by an angle between 0 and 2π. The connectivity is random (fig. 1 *left*) and the probability of connections, both inside and across the populations, is spatially modulated and given by:

$$P_A(\theta_i, \theta_j) = p_0^{(A)} + p_1^A \cos(\theta_i - \theta_j) \tag{1}$$

Here $A = E, I$ and refer either to "excitatory" (E) or inhibitory (I). $P_A(\theta_i, \theta_j)$ is the probability that a neuron in population A at angle θ_i establish a link with a neuron in any of the two populations and at angle θ_j. For simplicity the dependence on the receiving population has been dropped, and then the overall spatial profile of inhibitory and excitatory connections is set just by 4 parameters. By varying them, it is possible to generate random networks in which short-range connections are predominantly excitatory and long range connections predominantly inhibitory (mexican-hat connectivity) as well as random networks with inverterted mexican hat connectivity and all the intemediate cases. Different dynamical behaviors are obtained from the simulations in different regimes, including homogeneous activity states (fig. 2a), static and oscillating bumps of activity (fig. 2b, 2c), uniform oscillations (fig. refstatesd), standing waves (fig. 2e, 2f), chaos (fig. 2g, 2h).

It is possible to build a simple rate-model capturing the same richness of spatio-temporal activation patterns. The selected architecture is a single ring of N threshold-linear rate units and the connectivity is taken all-to-all (fig. 1 *right*). Once again a precise inhomogeneous profile of excitation and inhibition is introduced. The couplings between two units along the ring are indeed spatially modulated and assume the following form:

$$J(\theta_i, \theta_j) = J_0 + J_1 \cos(\theta_i - \theta_j) \tag{2}$$

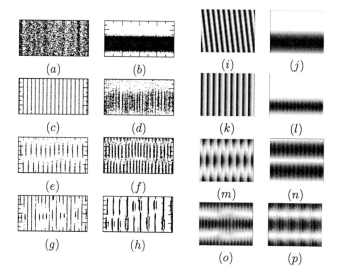

Figure 2: Typical firing patterns in the conductance-based network $(a-h)$ and in the corresponding delayed rate model $(i-p)$ [14, 15].

The evolution equations for the activity variables $m_1, \ldots, m_N$ associated to the rate units are:

$$\dot{m}_i(t) = -m_i(t) + \left[I_{ext} + \sum_{j=1}^{N} J(\theta_i, \theta_j) m_j(t - D) \right]_+ \tag{3}$$

where an interaction delay D has been explicitly taken into account. The simplicity of the resulting dynamical system makes possible to derive a phase-diagram which is in large part analytical (fig. 3, *left*). The homogeneous state instabilities can be computed and lead to a static bump state (SB, fig. 2j) via a Turing bifurcation, to uniform oscillations via a Hopf bifurcation (OU, fig.2l) and to traveling waves via a Turing-Hopf bifurcation (TW, fig.2i). A rate instability is also present and brings to epileptic states with unreasonably large activity values. Under certain assumptions on the relation between the delay D and the period of the oscillations, it is possible to build explicitly the limit cycle associated to uniform oscillations and to derive its instabilities with a Floquet analysis. The transition lines to standing waves states (SW, fig.2m and 2n) and to an oscillating bump state (OB, fig. 2k) can then be found, as well as the instability of the traveling wave which destabilizes first into "lurching waves" (LW) and then into standing waves. Interestingly, all these states are in one-to-one correspondence with the observed dynamical patterns of the conductance-based network, but now analytical insight about their origin has been gained. Further numerical explorations allow to identify regions of bistability among different states and irregular aperiodic collective oscillations (A, fig. 20 and 2p).

These last features cannot be understood analytically in the large N model. As a matter of fact, it turns out that most of the complex state-structure of the phase-diagram is preserved when the size is reduced to very small N values (in contrast with the ordinary statistical mechanics intuition, according to which finite-size effects could be able to alter completely the "thermodinamic limit" picture). Numerical experiments ascertain that already at $N = 4$ the phase-diagram is essentially

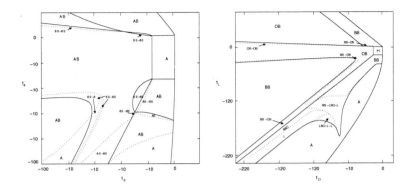

Figure 3: Left: phase diagram of the large N rate model. Right: phase diagram of the $N = 2$ rate model.

indistinguishable from the large N one. For smaller sizes some pathologies have to arise: at $N = 3$ for instance it becomes geometrically impossible to distinguish between standing and traveling wave states and some of the transition lines will disappear. At $N = 2$, the rate instability line moves considerably to the left and one of the two aperiodic regions disappears (fig.3, *right*). Still the homogeneous, the SB and the SB states are present, together with regions in which the two units oscillate regularly in phase (OU) or in antiphase (SW). We will focus in the following to the mechanisms explaining the generation of chaotic oscillations. When $N = 2$ it is more natural to rewrite eq. (3) in the following form:

$$\dot{m}_{1,2}(t) = -m_{1,2}(t) + \left[I_{ext} + K_0 m_{1,2}(t - D) + K_1 m_{2,1}(t - \bar{D}) \right]_+ \qquad (4)$$

where the new couplings are given by $K_0 = J_0 + J_1$ and $K_1 = J_0 - J_1$, and where we allow now the internal and the relative delays D and $\bar{D}$ delays to be different for generality.

A broad chaotic region is observed when the local interaction coupling K_0 is strongly negative. In fig. 4, the extent of the chaotic region is monitored by looking both at the largest Lyapunov exponent and at the second largest autocorrelation peak of the activity of the most irregular unit. When K_1 is slightly negative the network displays a bistability between two regularly periodic oscillating states, in phase or in antiphase (Fig. 5a and b). When $K_1 = 0$ the two units are perfectly decoupled. Still they continue to oscillate (the smallest oscillating "network" is indeed constituted by a single rate-unit with delayed self-inhibition), but their phase relation becomes completely arbitrary. As soon as a weak excitatory feedback connection is switched-on a new precisely phase-locked state is obtained, in which the phase-shift is intermediate between 0 and π (lag-synchronized oscillations, LSO, Fig. 5c). This *spontaneous symmetry breaking* allows to identify a neuronal unit ("leader") whose phase is systematically in advance with respect to the second unit ("laggard"). Similar transitions have been observed in other coupled nonlinear chaotic oscillator systems [21], like, for instance, delay-coupled semiconductor lasers [22]. Increasing now the coupling, a first period-doubling transition is observed (Fig. 5d), followed then by a Feigenbaum cascade of bifurcations (F; the scenario of the transition to chaos has been validated by looking at the convergence of the bifurcation diagram properties to the universal Feigenbaum constants).

Remarkably, in a first range of parameters the observed chaotic oscillations are strongly *asymmetric*. Both period-doubling and chaos manifest themselves mainly in (regular or irregular) fluctuations of the maxima of the oscillations, and the range of this amplitude variations is initially

101

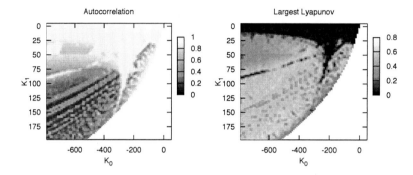

Figure 4: Autocorrelation and largest Lyapunov exponent in the $N = 2$ ring-model chaotic region.

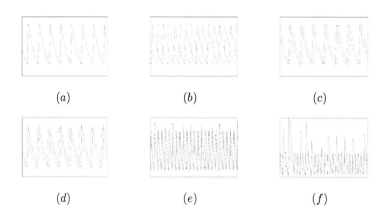

Figure 5: Traces of the $N = 2$ ring-model, $K_0 = -320$: (a) in-phase oscillations, $K_1 = -1$; (b) antiphase oscillations, $K_1 = -1$; (c) lag-synchronized oscillations, $K_1 = 1$; (d) period-doubled oscillations, $K_1 = 7$; (e) asymmetric chaos, $K_1 = 14$; (f) symmetric chaos, $K_1 = 50$

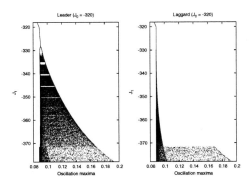

Figure 6: Bifurcation diagrams of the oscillation amplitudes for the leader and the laggard neuron.

considerably larger for the laggard neuron than for the leader (Fig.6). At a first visual inspection of the traces (Fig. 5e), it might seem that the leader neuron continues to oscillate periodically, but small fluctuations —two order of magnitude smaller relatively to the laggard neuron— are actually induced by the feedback connection. It is actually well known that biological oscillators can exhibit chaotic behavior when driven by sinusoidal sources or by pulsating inputs [23, 24, 25]. Thanks to the symmetry breaking, the much more regular signal of the leader neuron can be approximately treated as a periodic external current source driving the laggard unit. The hypothesis can be checked by excising one of the two excitatory reciprocal connections and creating then an explicitly asymmetric system. By varying K_1 at constant K_0 one obtains a bifurcation diagram for the oscillation amplitudes of the driven unit which is perfectly correspondent to the one of the laggard unit in the $N = 2$ ring model. Interestingly, the asymmetric driver-driven system undergoes a phase transition at a larger K_1 from chaotic oscillations to phase-locked 2:1 periodic oscillations. In the symmetric model, this phenomenon is replaced by a transition to a new symmetric chaotic phase (Fig.6). The increasing excitatory feedback induces then the generation of oscillations of growing amplitude, until when the rate instability line is crossed and an epileptic state of activity is entered.

Work is in progress in order to observe analogous symmetry-breaking mechanisms followed by transition to chaos in conductance-based models. The exact results obtained for the large N model can be easily extended to the $N = 2$ case. Furthermore, the simplicity of the latter should allow to prove rigorously the onset of symmetry-breaking and of the period-doubling route to chaos resorting to Kuramoto theory of weakly coupled oscillators and mean-field techniques.

Bibliography

[1] Ben-Yishai, R., Hansel, D., Sompolinsky, *J. Comp. Neurosci.* 4:57–79 (1997)

[2] Hansel D., Sompolinsky H., Modeling Feature Selectivity in Local Cortical Circuits, in *Methods in Neuronal Modeling*, edited by Koch C. and Segev I., MIT press, Cambridge, MA (1998)

[3] Golomb D., Ermentrout B., *Proc. Natl. Acad. Sci. USA* 96:13480(1999)

[4] Ben-Yishai R., Lev Bar-Or R., *Proc. Natl. Acad. Sci. USA* 92:3844 (1995)

[5] Somers D.C., Nelson S.B., Sur M., *J. Neurosci.* 15:5465–5488 (1995)

[6] Compte A., Brunel N., Goldman-Rakic P.S., Wang X.-J., *Cereb. Cortex* 10: 910–923 (2000)

[7] Gutkin B., Laing C.R., Colby C.L., Chow C.C., Ermentrout G.B., *J. Comput. Neurosci.* 1:121–134 (2001)

[8] Ermentrout B., *Rep.Prog.Phys* 61:353 (1998)

[9] Bressloff P.C., Pattern formation in visual cortex, in *Methods And Models In Neurophysics: Proceedings of the Les Houches Summer School 2003*, Elsevier (2004)

[10] Markram H., Lubke J., Frotscher M., Roth A., Sakmann B., *J. Physiol. (London)* 500:409 (1997)

[11] Fourcaud-Trocmé N., Hansel D., van Vreeswijk C., Brunel N., *J. Neurosci.* 17, 23:11628–40 (2003)

[12] Brunel N., Hakim V., *Neur. Comput.* 11:1621–1671 (1999)

[13] Brunel N., Wang X.-J., *Journal of Neurophysiology* 90:415–430 (2003)

[14] Roxin A., Brunel N., Hansel D., *Phys. Rev. Lett.* 94:238103 (2005)

[15] Roxin A., Brunel N., Hansel D., *Progr. Theor. Phys. Suppl.* 161:68–85 (2006)

[16] Hansel, D., Sompolinsky, H., *J. Comp. Neurosci.* 3:7–34 (1996)

[17] Van Vreeswijk, C.A., Sompolinsky, H., *Science* 274:1724–1726 (1996)

[18] Van Vreeswijk, C.A., Sompolinsky, H., *Neural Comp.* 10:1321–1372 (1998)

[19] Whittington M.A., Traub R.D., Jefferys J.G.R., *Nature* 373:612–615 (1995)

[20] Wang X.-J., Buzsáki G., *Journ. Neurosci.* 16:6402:6413 (1996)

[21] Rosenblum M.G., Pikovsky A.S., Kurths J., *Phys. Rev. Lett.* 78:4193 (1997)

[22] Heil T., Fischer I., Elssser W., Mulet J., Mirasso C.R., *Phys. Rev. Lett.* 86:795 (2001)

[23] Petrillo G.A., Glass L., *Am. J. Physiol.* 236, R311 (1984)

[24] Calabrese R.L., de Schutter E., *Trends Neurosci.* 15:439 (1992)

[25] Coombes S., Bressloff P.C., *Phys. Rev. E* 60:2086 (1999)

Ecological Networks

Ecological networks: methods and data

Jordi Bascompte
Estación Biológica de Doñana, CSIC
Apdo. 1056, E-41080, Spain

bf Abstract
In the last few years there has been a burst of interest in ecological networks. This has been the consequence of (i) the availability of a new generation of large datasets, and (ii) the borrowing of concepts, metaphors, and tools from the physics of complex networks. Essentially, recent work has focussed on two types of ecological networks. First, networks of species interactions, where the type of this interaction can be of different nature such as antagonistic (e.g., predator prey, host-parasitoid) or mutualistic (e.g., plants and their insect pollinators). Second, spatial networks, where two patches of discrete habitat are linked if a given species can move between them through direct dispersal. Here, I will briefly review the sort of data we have now, the methods employed in their analysis, and what it remains to be done for the next few years, with particular emphasis on what can be achieved from the UniNet experience.

Introduction

Ecology has always relied on a network framework (Margalef 1991). For example, food webs describing who-eats-whom have represented the complexities of ecological communities. Theoretical ecology has always looked at food web structure in the belief that such structure greatly affects community stability. Recently, however, concepts and tools from the physics of complex networks have been incorporated into the ecological agenda (Solé and Bascompte 2006). For a comprehensive review on food web research, with a particular emphasis on models, see the Chapter by Joan Saldaña.

Another field in ecology in which the network framework is providing a renewed interest is spatial ecology. The first theoretical models in population ecology were based on mean field equations describing the temporal dynamics of one or two-interacting species such as a prey and its predator. Space was missing. The incorporation of heterogeneous landscapes came later on, and in the last two decades spatial ecology has become a hot area in ecology, mainly because of the recognition than otherwise homogeneous landscapes are becoming fragmented due to human-induced habitat loss (Bascompte and Solé 1996, Hanski 1999). Thus, metapopulation research has become an important framework describing the persistence of species as a balance between colonization and extinction events (Hanski 1999). Disconnected habitat patches linked by dispersal can be described by a network approach. The paper by Urban and Keitt (2001) has been arguably the first serious attempt at bringing graph theory into spatial ecology. There is actually a renewed interest with several manuscripts circulating. Fortuna *et al.* (2006), for example, have used spatial networks to describe the persistence of amphibians in a set of fluctuating ponds.

Let's consider briefly the two applications of network theory and a description of data and methods in each one.

Ecological Interactions

Networks in here are defined by the following elements. Nodes are composed by ecological or taxonomic species (species who have the same set of prey and predators are oftentimes lumped together). Also, sometimes groups of similar species such as different insect species which can not be identified in the field are lumped together. A link is here defined as an interaction between two given species. In food webs, this is a trophic interaction, i.e., a predator consuming its prey. The level of noise in these networks is quite large and they are almost always a snapshot, lacking spatial and temporal variability. The new generation of ecological networks span from two dozens to two-hundred species and up to 3,000 interactions (e.g., Bascompte *et al.*, 2005). Earlier studies in food webs described aggregate measures such as connectance (fraction of all observed links), fraction of top predators, and similar (see Joan Saldaña's Chapter). More recently, studies of food webs describe topological properties such as degree distribution (Solé and Montoya 2001, Dunne *et al.*, 2002, Memmott *et al.*, 2004), clustering coefficient (Montoya and Solé 2002), connectivity correlation (Melián and Bascompte 2002), path length (Montoya and Solé 2002), and compartments. Compartments is an old concept already studied by Pimm and Lawton (1980) but that is now analyzed with techniques from social networks (Krause *et al.* 2003).

Another approximation to network structure has been that of motifs, i.e., patterns of inter-connections occurring in complex networks at numbers that are significantly higher than those in randomized networks (Milo *et al.* 2002). In ecology, these motifs are the biologically-meaningful trophic modules studied by theoreticians such as tri-trophic food chains or omnivory. Interestingly enough, the direction of research in ecology is here opposed to the one on complex networks: there were first theoretical studies on the dynamics of such trophic modules and only recently people have explored their relative frequency (Bascompte and Melián 2005, Bascompte *et al.* 2005). On the other hand, studies in complex networks have first identified network motifs (Milo *et al.* 2002), and only recently people have explored their dynamics (Prill *et al.* 2005).

In summary, this research has paralleled general research on complex networks, starting from connectivity distributions and going into deeper levels of structure (e.g., modularity, see Ravasz *et al.*, 2002).

Other ecological interactions besides predator-prey are mutually beneficial interactions such as the ones between plants and their animal pollinators or seed dispersers. These networks have been studied only in the past five years (see however Jordano 1987). They are, on average, better resolved and contain a larger number of species, and are represented by bipartite graphs, i.e., graphs with two sets (plants and animals), with interactions between but not within sets (Jordano *et al.*, 2003; Bascompte *et al.*, 2003; 2006).

Mutualistic interactions are very important because they are acknowledged as an important mechanism in generating biodiversity on Earth. The origin of pollinating insects opened new niches for flowering plants, which in turn spurred the diversification of insects (Ehrlich and Raven 1964, Thompson 2005). For example, more than 90 % of tree species in tropical forest depend on animals for seed dispersal, and thus would cease to exist without these animals. These mutualistic inter-actions have been mainly studied as pairwise specific interactions, and only recently the network approach has allowed to consider the whole community.

More than 50 plant-animal mutualistic communities have been compiled and analyzed, which gives an unusual level of generality to these results. In this case, studies have looked at degree distribution (Jordano *et al.*, 2003), its quantitative extension, species strength (Bascompte *et al.*, 2006), nestedness (Bascompte *et al.*, 2003), clustering coefficient and path length (Olesen *et al.*, 2006), and asymmetry in mutual dependences between plants and animals (Bascompte *et al.*, 2006). Analysis of ecological interactions have been mainly descriptive. For example, people has been interested in describing the shape of the connectivity distribution echoing work by Albert *et al.* (2000) relating connectivity distribution to the robustness of complex networks. Although a few food webs have power-law connectivity distributions, the bulk of them are exponential (Solé and Montoya 2001, Camacho *et al.* 2002, Dunne *et al.*, 2002). Interestingly enough, mutualistic networks are mainly described by broad scale distributions (i.e., truncated power-laws). Both Solé and Montoya (2001), Dunne *et al.* (2002) and Memmott *et al.* (2004) have simulated node removals to stress the level of robustness of these networks, although topological robustness to node removal was much earlier studied by Pimm (1980).

Spatial Networks

As stated above, the other area where network research has contributed to ecology is spatial dynamics. Urban and Keitt (2001) used graph analysis to explore the connectivity of habitat patches in a nice example of applicability to the conservation of the spotted owl. However, very few papers (e.g., Cantwell and Forman 1993; Fagan 2002) have followed this path although this is now a hot area and several papers are in press. A network is now defined by a collection of habitat patches, e.g., ponds or forest patches. These are separated by a matrix of inhospitable habitat (e.g., agricultural fields). Two patches are linked if a target organism can disperse from one to the other. In the largest spatial network analyzed within this framework, there are more than 3,000 nodes. The number of links depends on the dispersal distance of each species, but may be in the order of several thousands. Similar variables than the ones analyzed in interaction networks have been applied, mainly degree distribution, and clustering coefficient. Fortuna *et al.* (2006) detected changes in network topology of temporary ponds as the level of drought increased. Temporary ponds in Mediterranean regions such as Doñana National Park, where the study was carried out, are highly variable. Fortuna *et al.* (2006) simulated such variability by increasing the fraction of ponds that dry out. This is an interesting extension of Albert *et al.* (2000) node removal, but in here nodes are removed on the basis of their size (this information, as well as the spatial coordinates were provided by GIS). Interestingly enough, the topology of the network confers a high degree of cohesion even for large drought values. Thus, despite deleting ponds, the remaining ones constitute a good refuge for amphibians.

Analysis of network topology provide a new and straightforward way to quantify the robustness of a patchy population to habitat loss and the identification of keystone patches that are critical to landscape connectivity and hence population persistence (Urban and Keitt 2001; Keitt 2003). The application of existing network approaches to spatial processes can yield valuable insight. Network analysis is a powerful tool for analyzing real landscapes and can also be used as a basis for building more complex population viability models.

Prospects

There is now a new generation of large, high quality data sets of ecological networks, specifically interaction networks (both food webs and mutualistic networks), and spatial networks. Tools and concepts from the physics of complex networks have been imported and recent effort has described the architecture of such networks with important implications for community stability, conservation, and coevolution (see Proulx *et al.* 2006, May 2006, and Montoya *et al.* 2006 for recent reviews).

Besides the papers exploring topological robustness, studies of ecological networks have been mainly descriptive. In a study of a very large food web in the Caribbean, Bascompte *et al.* (2005) have linked basic weighted network motifs to population dynamics by means of a bioenergetic model. This type of experience allows to explore the community-wide consequences of overfishing. However, this study relies only on simple trophic modules, uncoupled from the rest of the community. The challenge is to build dynamic models incorporating the exact topology of full interacting networks. A first example is by Fortuna and Bascompte (2006) who used real plant-animal mutualistic networks and appropriate network randomizations to build several dynamic models. Comparing the outputs of these models, one can explore the role of network structure. In this example, a spatial model was used and the role was to explore the response to habitat fragmentation. Due to the highest heterogeneity in degree for the real network, and their nested structure, real communities start to loss species sooner than randomizations (generalist go extinct first), but stand until larger values of destruction (Fortuna and Bascompte 2006). There are lots of opportunities here to improve these models, extend them to other scenarios, and build analytic approximations to study what is the consequence of network architecture for community dynamics. Here I identify an area with potential interaction within UniNet. Other UniNet nodes are experts on analytical approaches to characterize the stability of full networks. Bringing this expertise to our ecology node would allow to characterize the stability of communities.

Bibliography

[1] Albert, R., Jeong, H., and Barabási, A.-L. (2000). Error and attack tolerance of complex networks. *Nature* **406**: 378-382.

[2] Bascompte, J., Melián, C.J., and Sala, E. (2005). Interaction strength combinations and the overfishing of a marine food web. *Proc. Natl. Acad. Sci. USA* **102**: 5443-5447.

[3] Bascompte, J., Jordano, P., Melián, C.J., and Olesen, J.M. 2003. The nested assembly of plant-animal mutualistic networks.*Proc. Natl. Acad. Sci. USA* **100**: 9383-9387.

[4] Bascompte, J., Jordano, P. and Olesen, J.M. 2006. Asymmetric coevolutionary networks facilitate biodiversity maintenance. *Science* **312**: 431-433.

[5] Bascompte, J., and Melián, C.J. 2005. Simple trophic modules for complex food webs. *Ecology* **86**: 2868-2873.

[6] Bascompte, J. and Solé, R.V. 1996. Habitat fragmentation and extinction thresholds in spatially explicit models. *J. Anim. Ecol.* **65**: 465-473.

[7] Camacho, J., Guimerà, R., and Amaral, L.A. 2002. Robust patterns in food web structure. *Phys. Rev. Lett.* **88**: 228102.

[8] Cantwell, M.D., and Forman, R.T.T. 1993. Landscape graphs: ecological modeling with graph theory to detect configurations common to diverse landscapes. *Landscape Ecol.* **8**: 239-251.

[9] Dunne, J., Williams, R.J., and Martinez, N.D. 2002. Network structure and biodiversity loss in food webs: robustness increases with connectance. *Ecol. Lett.* **5**: 558-567.

[10] Dunne, J.A., Williams, R.J., and Martinez, N.D. 2002. Food-web structure and network theory: the role of connectance and size. *Proc. Natl. Acad. Sci. USA* **99**: 12917-12922.

[11] Ehrlich, P.R., and Raven, P.H. 1964. Butterflies and plants: a study on coevolution. *Evolution* **18**, 586-608.

[12] Fagan, W.F. 2002. Connectivity, fragmentation, and extinction risk in dendritic metapopulations. *Ecology* **83**: 3243-3249.

[13] Fortuna, M.A., and Bascompte, J. 2006. Habitat loss and the structure of plant-animal mutualistic networks. *Ecol. Lett.* **9**: 281-286.

[14] Fortuna, M.A., Gómez-Rodríguez, C., and Bascompte, J. 2006. Spatial network structure and amphibian persistence in stochastic environments. *Proc. R. Soc. London B* **273**: 1429-1434.

[15] Hanski, I. 1999. *Metapopulation Ecology*. Oxford University Press.

[16] Jordano, P. 1987. Patterns of mutualistic interactions in pollination and seed dispersal: connectance, dependence asymmetries, and coevolution. *Am. Nat.* **129**: 657-677.

[17] Jordano, P., Bascompte, J., and Olesen, J.M. 2003. Invariant properties in coevolutionary networks of plant-animal interactions. *Ecol. Lett.* **6**: 69-81.

[18] Keitt, T.H. 2003. Network theory: an evolving approach to landscape conservation. In *Ecological Modeling for Resource Management*, V.H. Dale, ed. pp. 125-134, Springer, New York.

[19] Krause, A.E., Frank, K.A., Mason, D.M., Ulanowicz, R.E., and Taylor, W.W. 2003. Compartments revealed in food-web structure. *Nature* **426**: 282-285.

[20] Margalef, R. 1991. Networks in Ecology. In *Theoretical Studies of Ecosystems- The Network Perspective*, M. Higashi and T.P. Burns, eds. pp. 41-57, Cambridge University Press.

[21] May, R.M. 2006. Network structure and the biology of populations. *Trends Ecol. Evol* **20**: 345-353.

[22] Melián, C.J., and Bascompte, J. 2002. Complex networks: two ways to be robust? *Ecol. Lett.* **5**: 705-708.

[23] Memmott,J., Waser, N.M. and Price, M.V. 2004. Tolerance to pollination networks to species extinctions. *Proc. R. Soc. London B* **271**: 2605-2611.

[24] Milo, R., Shen-Orr, S., Itzkovitz, S., Kashtan, N., Chklovskii, D., and Alton, U. Network motifs: simple building blocks of complex networks. *Science* **298**: 824-827.

[25] Montoya, J.M., and Solé, R.V. 2002. Small world pattern in food webs. *J. theor. Biol.* **214**: 405-412.

[26] Montoya, J.M., Pimm, S.L., and Solé, R.V. 2006. Ecological networks and their fragility. *Naure,* in press.

[27] Olesen, J.M., Bascompte, J., Dupont, Y.L., and Jordano, P. 2006. The smallest of all worlds: pollination networks. *J. theor. Biol.* **240**: 270-276.

[28] Pimm, S.L. 1980. Species deletion and the design of food webs. *Oikos* **35**: 139-149.

[29] Pimm, S.L. and Lawton, J.H. 1980. Are food webs divided into compartments? *J. Anim. Ecol.* **49**: 879-898.

[30] Prill, R, Iglesias, P. Levchenko, A. 2005. Dynamic properties of network motifs contribute to biological network organization. *PLoS Biology* **3**:e343.

[31] Proulx, S.R., Promislow, D.E.L., and Phillips, P.C. 2006. Network thinking in ecology and evolution. *Trends Ecol. Evol* **20**: 345-353.

[32] Ravasz, E., Somera, A.L., Mongru, D.A., Oltavi, Z.N., and Barabási, A.-L. 2002. Hierarchical organization of modularity in metabolic networks. *Science* **297**: 1551-1555.

[33] Solé, R.V. and J.M. Montoya. 2001. Complexity and fragility in ecological networks. *Proceedings of the Royal Society of London B* **268**: 2039-2045.

[34] Solé, R.V. and J. Bascompte. 2006. *Self-Organization in Complex Ecosystems*. Princeton University Press.

[35] Thompson, J.N. 2005. *The geographic mosaic of coevolution*. Chicago University Press.

[36] Urban, D. and Keitt, T. 2001. Landscape connectivity: a graph-theoretic perspective. *Ecology* **82**: 1205-1218.

Overview: Modelling complex food webs

Joan Saldaña
Departament d'Informàtica i Matemàtica Aplicada
Universitat de Girona, 17071 - Girona, Spain

Abstract
After a brief survey of the statistical properties and of the main stochastic models of static food webs, recent models integrating structure and dynamics are discussed.

Introduction

Ecological systems consist of an assembly of animals and plants interacting in different ways as those defined by predator-prey relationships, mutualism, antagonism, competition by resources, parasitism, etc. However, due to the difficulties of collecting the suite of data necessary to give a complete enough picture of such interactions, field ecologists have traditionally focus their research on single types of them, mainly on those related with feeding relationships which define the so-called *food webs* . The consequence of this is that, according to the sort of interaction we are interested in, a particular description of an ecological system arises (see [2, 3] for a recent description of the structure of animal-plant mutualistic networks and [16] for the statistical description of 16 high-quality food webs).

Although in most of the cases, the interacting species live in the same physical location, sometimes an ecological system under study includes a geographical collection of local populations of different species connected by migratory flows of their individuals. Such collections are called *metapopulations* and their consideration in the description of the ecological system implies an additional degree of complexity. Only few and recent studies have attempted to link structure and dynamics in these spatially structured communities of several species (see, for instance, [1, 26]).

The first descriptions of ecological systems were in terms of "who eats whom" relationships among biological species and they go back to the late 1800s (Forbes, 1876; Camerano, 1880). By the 1920s the first relatively detailed empirical descriptions of terrestrial and marine food webs appeared (see [17] for a historical review of food web theory). These initial studies were not very extensive and probably they were made with a biased recording towards the more abundant species. After these early descriptions, food webs with an increasing level of resolution have been reported. The first published collection of food webs appeared in 1978 and it was due to Joel E. Cohen who compiled 30 food webs from the literature. The same author joint with F. Briand, and C.M. Newman published in 1990 a review ([11]) of 113 webs where the number of species vary from 5 to

48. Since then, more comprehensive food webs have been reported some of them with more than 100 species. For instance, one of the species-richest food web described up to know, the one of the Caribbean Coral Reef, contains more than two hundred trophic species. A *trophic species* is defined as "one or more taxonomic species that eat entirely the same set of prey and are eaten by an entirely identical set of predators" ([10]).

Statistical description of food webs

A food web consists of a set of species (nodes) and a set of trophic relationships (links) among them. Therefore, the most simple description of a food web is the one given by a binary (or non-weighted) graph and, hence, with two basic quantities associated with it: the number S of species in the web, and the total number of trophic links L. These quantities define the most important measures in the web, namely, the *linkage density* given by the ratio L/S, the so-called *directed connectance* $C = L/S^2$ (the fraction of all possible links that are realized) used when considering food webs as directed graphs, and an analogous measure for undirected graphs, the *interactive connectance* $C_u = 2L/(S(S-1))$, which excludes mutual predation and intraspecific interactions ([24]).

The early works were mainly oriented to find empirical regularities of the topological description of food webs, sometimes called food-web patterns. In particular, those patterns "in which a property is found either to be constant, or under a weaker standard to not change systematically as the number of species across food webs changes, came to be referred to as *scale-invariant patterns* or *scaling laws* " ([17]). The measures whose behaviour has been analyzed were the linkage density and the directed connectance. As a conclusion, it seems clear that the scale invariance hypothesis for the linkage density (i.e., $L \approx aS$), initially proposed by J.E. Cohen in [10] from the analysis of the first catalog of food webs, is not supported by high-quality, species-rich food webs recently reported and that no consensus has emerged to replace it ([14]). From the analysis of larger food webs, Martinez proposed in 1992 the hypothesis of constant connectance, which implies $L \approx aS^2$, which has been also called into question by later studies ([17]).

Other attributes that have been also traditionally used to describe food-web patterns are the proportion of top (T), intermediate (I), and basal (B) species, and in terms of the proportion of links among these classes (II, TI, IB, TB). Top species are species with no predators while basal species are those with no prey. For all these quantities similar scale invariance hypotheses were proposed in [11]. However, from the major reviews of food web patterns reported in the 1990s and from the re-analysis of their patterns under different species aggregation criteria, it became clear that there was no empirical evidence that food-web patterns changed with S ([14]).

More recently, the topology of food webs has been analyzed by means of descriptors used in complex networks as the characteristic path length (the average shortest distance between pairs of nodes), the connectivity distribution (distribution of the number of (trophic) links per node (species)), clustering coefficient (the average fraction of pairs of nodes connected to the same node that are also connected to each other), or degree correlations (see [32]). As complex networks, food webs have relatively few nodes compared with non-biological networks and clustering coefficients that are much closer to those of randomly generated networks ([16]). Moreover, in their topological description, the reported connectivity distributions are systematically related to food-web connectance and number of species. More precisely, from the currently available data it is recognized that, although some food webs show the so-called "small-world" property (i.e., short paths between nodes and high clustering) and a connectivity distribution given by a power law, most do not if they exceed a relatively low level of connectance. In turn, the small size of food webs compared with most of other networks appears to be also related to this deviation of the majority of food webs from the small-world topology. In fact, when network size is considered, food webs fit into a predictable continuum of clustering, with increasingly greater than random clustering observed in larger networks (see [32] for details). What it seems to be a new formulation of a universal law is the functional form of some descriptors like clustering coefficient or characteristic path length, that scales with linkage density after the size of network and the distribution of the number of links are scaled by the average network connectivity $2L/S$ ([7]).

A quantitative description of ecological networks is an alternative modelling approach to an analysis focused on the network structure. Under this approach, links are no longer binary, but weighted. They are quantified by means of different measures depending on the nature of the ecological network as, for instance, by the frequency of visits in plant-pollinator mutualistic networks ([2]), or by the energy (or biomass) flow along trophic interactions in food webs. Moreover, also nodes can be quantified by focusing on species traits, like body mass, and numerical abundances ([12]). The picture we finally obtain is, therefore, a static food web with weights attached to nodes (species biomass) and/or links (interaction strength). However, it remains argued whether these as well as other approaches applied to food webs will constitute a basic tool to describe and identify new patterns as it seems to be the case according to [12] (see [17] for a discussion).

Stochastic models for static food webs

In order to reproduce the topological patterns observed in empirical food webs, three main stochastic models have been developed for *static network topologies*, i.e., for webs with a fixed number of nodes and binary (unweighted) links. The two first models are based on the empirical observation that most of food web are *interval*, which means that "all species in a food web can be placed in a fixed order on a line such that each predator's set of prey forms a single contiguous segment of that line" ([17]). This clearly indicates the one-dimensionality of the niche space and constitutes the starting point for several model formulations. In all cases, the generation of food-web structure is based on the creation of a feeding hierarchy obtained by assigning a random value x_i (*niche value*) to each species drawn uniformly from $[0, 1]$. This value x_i determines the rank of the i-species in the feeding hierarchy and, hence, which species can eaten by i-species, i.e., the i-species' feeding range.

In the so-called *cascade model*, the first relevant stochastic model for food webs ([10]), the i-species' feeding range is simply given by species with niche values $x_j \in [0, x_i)$. The model is tuned both in web size S and linkage density L/S, to match the observed data. The probability of feeding on species with niche values within this range is constant and equal to $2CS/(S-1)$.

More recently, it was proposed the so-called *niche model* ([33]), probably the most successful stochastic model for static food webs. This model is tuned both in web size S and connectance C, to match the observed data. In this model, the i-species' feeding range has a size $r_i = a_i x_i$ where $a_i \in [0, 1]$ is assigned by using a beta function (with $\alpha = 1$) whose expected value is $2C$. Therefore, the resulting connectance of the model web is close to C, the target value. Finally, the center c_i of the feeding range is drawn uniformly from $[r_i/2, x_i]$ if $x_i \leq 1 - r_i/2$, or from $[r_i/2, 1 - r_i/2]$ otherwise. This choice guarantees that c_i is always less than or equal to the i-species niche value x_i and relaxes the cascade hierarchy by permitting cannibalism and feeding on species with higher niche values.

One of the last stochastic models devised for the description of food-web structure was proposed in 2004 by Cattin *et al.* The model is built on the hypothesis that "any species' diet is the consequence of phylogenetic constraints and adaptation" ([8]). The aim is to generate a network where consumer species are organized in groups forming a nested hierarchy. This is achieved by imposing a sequence in the attribution of links, which is partially inspired by Sugihara's niche-hierarchy model for generating species' relative abundances ([31]). The order of this sequence of attributions is given by the feeding hierarchy in such a way that prey are first attributed to primary (lowest ranked) consumers. Moreover, as in the niche model, there is an initial assignment of feeding ranges r_i to determine the the number of prey species for species i, namely, $r_i S$. The nested hierarchy arises because, after an assignment of a prey j to a given consumer species, the pool of potential prey of this consumer for the next assignment increases with the set of prey of those consumers such that: (a) share at least one prey species, and (b) at least one of them feeds on species j (see ([8]) for more details about the construction of consumer nested groups).

Several papers discussing the suitability of the predictions of the previous (and other) models have been published. However, under UniNet approach to complex systems, the more illuminating papers are [6, 30]. These two papers show for the first time analytical results for large model webs (with sparse interaction matrices) generated by using a general family of stochastic models. Their main result is the existence of two sufficient conditions for network models to reproduce the distribution of (scaled) number of prey, predators, and links of currently available food-web data ([7]). These conditions are (1) the niches values to which species are assigned form a totally ordered set, and (2) each species has a specific probability, drawn from an approximately exponential distribution, of feeding on species with lower niche values. Clearly, the niche and the nested-hierarchy models meet both conditions, while the cascade model does not satisfies the second one. However, this model can be easily generalized to satisfy the latter by assuming a species-specific feeding probability drawn either from the beta distribution (with $\alpha = 1$) or from an exponential distribution (see [30]). In all these three models, the generated distributions of number of prey, predators and links are the same after they are scaled by the average connectivity of the network, $2L/S$.

Dynamical models of food webs

The classical approach in population dynamics to the study of mutualistic and/or trophic networks is based on the use of systems of differential equations of Lotka-Volterra type for species abundances, with constant coefficients defining the set of interactions among species. The web structure as well as the interaction strengths are, therefore, fixed beforehand. The resulting picture is that of a network with a static architecture and temporal changes in the weights of its nodes.

The first attempt of integrating dynamics with complex food-web topology was due to R.M. May ([24]). In this work, a randomly configured network, without any temporal variation in its link structure, was the starting point to obtain the first analytical results on stability and complexity for this type of systems (see, for instance, the classical book by R.M. May [25]). Such an assumption is clearly restrictive but allows the use of analytical techniques devised for dealing with large random matrices. More recently, random configurations have been also used in numerical simulations of the same model to call into question May's famous claim that too large a connectance (or too strong an average interaction strength) leads to instability of randomly generated complex systems ([19]).

There are three reasons, at least, for which this approach to modelling dynamical food webs is unsatisfactory. First, randomly configured networks may allow unrealistic food-web architectures such as those without basal species ([9]). Secondly, the distribution of interaction strengths among species are likely to be non-constant since, in general, they should change in response to changes in species abundances. Finally, in Lotka-Volterra equations it is assumed non-saturating linear functional responses (foraging/consumption rates per unit of predator biomass).

¿From the early 2000s, some papers have addressed the first of these flaws by incorporating a more realistic food web architecture according to some of the stochastic models reviewed in the previous section. For instance, Chen and Cohen in [9] considered a dynamics governed by Lotka-Volterra equations for food webs with an architecture generated by the cascade model (see also [20]). Later models dealt with food-web topologies generated with the niche model in conjunction

115

with saturating nonlinearities as the so-called Holling type II functional responses, which assumes a saturation in the consumption rates as that of Michaelis-Menten equation for rate of enzyme mediated reactions (see, for instance, [5, 21, 29]). Other recent papers analyze the influence of the population dynamics on the long-term evolution of food-web structure ([13, 15]). A survey on functional responses is given in the review by Drossel and McKane [14].

A frequently used model is the one corresponding to a multispecies generalization of a *bioenergetic model* of trophic interactions initially proposed by Yodzis and Innes in 1992. Concretely, in a *n*-species community with biomass and carrying capacity given by N_i and K_i $(i = 1, \dots, n)$, respectively, the model reads

$$\frac{dN_i}{dt} = r_i N_i \left(1 - \frac{N_i}{K_i}\right) - l_i N_i + \sum_{j=1}^{n} \left(l_i y_{ij} \alpha_{ij} F_{ij}(N) N_i - l_j y_{ji} \alpha_{ji} F_{ji}(N) N_j \right) \tag{1}$$

where r_i is the intrinsic growth and is nonzero only for basal species, l_j is the mass-specific metabolic rate, y_{ij} is the maximal rate at which predator i assimilates prey j, and $\alpha_{ij} \geq 0$ is the *relative capture (or foraging) effort* of species i allocated to species j, which satisfies $\sum_{j=1}^{n} \alpha_{ij} = 1 \,\forall\, i$. A simple example of α_{ij} is the fraction of time invested in obtaining (i.e., searching and preying) a biomass unit of prey j by species i. F_{ij} is the foraging rate of prey j per predator i (when all the foraging efforts of this predator are allocated to that prey). According to the Holling type II functional response, one has

$$F_{ij}(N) = \frac{N_j}{\sum_{k=1}^{n} \alpha_{ik} N_k + N_{ji}^0} \tag{2}$$

where N_{ji}^0 is the half saturation density of species j when consumed by species i.

This kind of models nicely couples dynamics and structure since the set of efforts $\{\alpha_{ij}\}$ defines the weighted graph representing the food-web configuration. If, moreover, we add time evolution rules for these efforts, as those recently introduced by Kondoh ([20, 21]), this results in an evolving network with temporal changes in the weights of its links and nodes, that is, in a *dynamical system defined on a graph*. Hence, we have overcome the last remaining modelling flaw of those mentioned before.

An example of the differential equations for the efforts appears in [20] and is given by

$$\frac{d\alpha_{ij}}{dt} = g_i \alpha_{ij} \left(l_i y_{ij} F_{ij}(N) - \sum_{k=1}^{n} l_i y_{ik} \alpha_{ik} F_{ik}(N) \right) \tag{3}$$

where g_i is the so-called adaptation rate of species i. This equation simply says that the consumption preference of an adaptive predator i (i.e., with $g_i > 0$) for a given species j increases if the energy gain from this prey per unit effort (when all the predator consumption is focused on this prey) is higher than the average energy gain available with the current assignment of efforts $\{\alpha_{ij}\}$.

Other authors have also considered temporal changes in the weights of links but without considering a particular dynamics for these changes (see, for instance, [13, 22]). This can be done by assuming two different time scales, namely, a slow dynamics for populations (nodes) and a (very) fast dynamics for the diet choice (links). In fact, the diet-choice dynamics is so fast that adaptive predators are assumed to be omniscient with respect to prey abundances and perfect optimizers. In this sense, the optimal predator diet choice is given by an allocation of efforts that leads to an equal energy gain per unit effort for all prey ([13]) which corresponds to the so-called "ideal free distribution" of predators ([22]). In any case, whatever of both approaches it is used for the effort dynamics, the resulting network is endowed with a flexible architecture that allows for a richer species composition than non-flexible networks ([23]). This fact, indeed, has been recently supported by empirical evidence obtained from marine communities ([27]).

All these works, as well as others, show the deep interplay that takes place between structure and dynamics in complex food webs: structure acts on dynamics and, vice versa, dynamics acts on structure ([23]). Fortunately for UniNet project, a deep understanding of the implications of simple trophic modules (like those recently analyzed in [4]) on the dynamics and topology of a complex food web is still missing. Similarly, analytical results on these dynamical systems defined on graphs are almost non-existent and, as far as we know, focused on conditions for stability and permanence in the Lotka-Volterra cascade model with a large number of species ([9]).

References in the end of the following chapter.

Food webs as complex adaptive networks

Josep L. Garcia-Domingo
Universitat de Girona
Dept. Informàtica i Matemàtica Aplicada
PIV- Campus Montilivi
E-17071 Girona, Spain

Abstract
Food webs have been mathematically modelled as dynamical systems defined on a graph given by
the adjacency matrix of the interactions between species. These interactions have been considered
constants through the evolution of the web. We analyze some recent models where interactions
fluctuate in time, giving rise to more realistic (and complex) food web models.

Introduction

There exist different mathematical approaches for modelling food-webs. Most of them uses a matrix
to express the trophic interactions (predation, mutualism, competition) between the species in a
food-web. This matrix can be seen as the adjacency matrix of a (directed or undirected multi-
)graph, where vertices represent the different species (or trophic species) and edges (or links)
represent the existence of interactions between species. If the matrix is weighted, each entry
encode the type and the strength of the interaction.

The evolution in time of the food-web is modelled as a dynamical system. Perhaps the Lotka-
Volterra (LV) is the most popular model of this type. If we denote by $x_i = x_i(t)$ the density (or
biomass) of species i, the LV model reads as:

$$\frac{dx_i}{dt} = x_i(b_i + \sum_{j=1}^{N} a_{ij}x_j), \quad i = 1\ldots N,$$

or compactly written

$$\frac{dx_i}{dt} = x_i(b_i + (AX)_i), \quad i = 1\ldots N,$$

where $A = (a_{ij})$ is the interaction matrix, X the vector of densities, N the total number of species
and $(M)_i$ stands for the i-row of the matrix or vector M. Here, the sign of a_{ij} determines the type
of interaction between species i and j: if $a_{ij} > 0$, species j enhances growth of species i (predation
if i feeds on j or mutualism if i benefits from j), while $a_{ij} < 0$ says that j diminishes growth of i

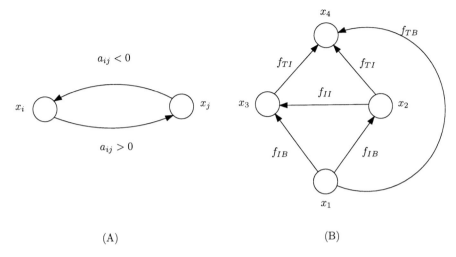

(A)   (B)

Figure 1: (A) Graph representation of possible interactions between i and j (the reverse a_{ji} not considered.) (B) A schedule of a trophic module of four species (one top, two intermediate and one basal) with omnivory (f_{TB}) and intraguild predation (f_{II}).

(predation if j feeds on i or competition if i and j compete for a recourse). Moreover, $|a_{ij}|$ is the strength of the interaction. If $a_{ij} = a_{ji} = 0$ there is no link between i and j. The b_i stands for birth rate of i if it is positive and for death rate if it is negative.

We can think that **the dynamical system is defined on a graph** given by the interaction matrix A. It is plausible to believe that the topology of the given graph (loops, connectance,...) can influence the flow of the dynamical system. Observe that also the evolution of the system can modify the graph: if $x_i(T) = 0$ for some $T > 0$ (and thus $x_i(t) = 0$ for all $t > T$), then the vertex corresponding to species i disappears as well as every link emanating from/arriving to i. Thus, **the dynamical system acts on the graph.**

In the LV model (and mostly in its subsequent generalizations), the interaction strengths (a_{ij}) do not vary in time, they are taken constant. This means that every species interacts with other given species with fixed intensities, everything independent of time. More realistic models will need to consider time-dependent interactions, due to behavioral adaptation of species or evolutionary changes. In this way, fluctuations in the intensities of predation must be allowed, but also switches in food choices. These assumptions give rise to more flexible (graph) structures and dynamics.

Although there has been multiple mathematical results analyzing non-autonomous dynamical systems with few species, we focus our attention on two recent approaches of modelling large food webs with time-dependent interactions. They can be catalogued as *foraging adaptation* or *adaptive predator* models.

- **Instant Adaptation Model (IAM).** This model is considered in [22], [13] and [15]. The formulation of these dynamical models are not coincident (basal species are treated different and different functional responses are considered), but they share the same hypothesis and method: essentially, the predator-prey interaction strengths are measured in terms of the predator total foraging effort (or time) distributed among its preys. The predators are assumed to be omniscient and the distribution of efforts instantly (for every $t > 0$) maximizes their total rate of gain in consuming preys (their *fitness*).

- **Dynamic Adaptation Model (DAM).** In this model, considered by [20],[21] and [5], the foraging efforts (that, as before, essentially establishes the predator-prey interaction strength) of every predator respect each of its preys is not updated instantly to maximize some predator fitness but it is driven by a differential equation of replicator type that compares the gain in consuming a particular prey with the average gain over all the predator preys.

This point of view formally involves that this new dynamics relies not only on the vertices but also on the edges, and consequently the action of the dynamics on the graph can now make disappear links in addition to nodes, thus obtaining a richer spectrum of resulting graphs. Our purpose is to analyze the results obtained in [20],[21] and [5] with the DAM paying attention on the final topological configuration of the food web, emphasizing the role played by the foraging adaption.

Methods and Models

We introduce the DAM and postpone the discussion of similarities and differences between the DAM and the IAM.

We use a bioenergetic dynamics ([5]). If $x_i := x_i(t)$ is the biomass of species i and X the biomass vector of all species, its dynamic equation is

$$\frac{dx_i}{dt} = r_i\left(1 - \frac{x_i}{k_i}\right) - l_i x_i + \sum_{j=1}^{N} l_i m_{ij} F_{ij}(X) a_{ij} x_i - \sum_{j=1}^{N} l_j m_{ji} F_{ji}(X) a_{ji} x_j, \tag{1}$$

where

$$F_{ij}(X) = \frac{f_{ij} x_j}{B_{ji}^0 + \sum_{k=1}^{N} a_{ik} x_k}$$

is the Holling's type II functional response and $a_{ij} := a_{ij}(t)$ is the rate of foraging effort that predator i spends on prey j, which dynamics is given by

$$\frac{da_{ij}}{dt} = g_i a_{ij}\left(l_i m_{ij} F_{ij}(X) - \sum_{j=1}^{N} l_j m_{ji} a_{ji} F_{ji}(X)\right), \tag{2}$$

and parameters are: r_i the growth rate; k_i the carrying capacity ; m_{ij} the maximum assimilation rate of i in consuming j; f_{ij} the foraging efficiency of i in consuming j; B_{ji}^0 the half-saturation rate of i in consuming j; g_i the adaption rate of i; l_i the mass-specific metabolic rate. The values used in our simulations are given in Table 1. Since our intention is to analyze the model from a theoretical point of view, the choice of the parameter values do not take into consideration biological meanings.

The efforts of all predator i satisfy the trade-off convention

$$\sum_{j=1}^{N} a_{ij} = 1 \qquad \forall t > 0. \tag{3}$$

This is guaranteed by

$$\sum_{j=1}^{N} \frac{da_{ij}}{dt} = 0,$$

Table 1: Parameters values used in simulations A, B and C.

parameter	A	B	C
r_i	1 (basals) 0 (non-basals)	1 (basals) 0 (non-basals)	1 (basals) 0 (non-basals)
k_i	1	1	1
m_{ij}	6	1	1
B_{ji}^0	1	1	1
g_i	1	1	1
f_{ij}	1	$f_{TB}=4, f_{TI}=12$ $f_{IB}=4, f_{II}=10$	$f_{TB}=4, f_{TI}=12$ $f_{IB}=6, f_{II}=18$
l_i	0.5	$l_B=0.15$ $l_T=l_I=0.5$	$l_B=0.15$ $l_T=l_I=0.5$

derived from equation (2), and by the correct choice of the initial conditions $a_{ij}(0)$ satisfying (3) at time $t = 0$. The $a_{ij}(t)$ functions give the structure of the underlying graph of the food web. Note that $a_{ij}(T) = a_{ji}(T) = 0$ for some $T \geq 0$ (and hence for all $t \geq T$) means that the graph at time T doesn't have a link between i and j (and nevermore). That $a_{ij}(t) > 0$ indicates that i is a predator of j at time t, and thus the graph has a link between i and j at that time.

In our numeric simulations, the starting food web structure is given by two different patterns: the *cascade* model and the *niche* model. We generate a random matrix with one of these models and we proceed to assign $r_i = 1$ for basal species and $r_i = 0$ for non-basal species, and the initial values for a_{ij} are taken $a_{ij}(0) = 1/(\text{initial number of preys of } i)$. The initial values for densities $x_i(0)$ are randomly taken between 0 and 1.

Since we want to study the influence of foraging adaption in the food web, next step is to assign the ability of being an adaptive predator or not. In [20],[21] and [5], it is done randomly: a species is capable of adaptive foraging with probability F $(0 \leq F \leq 1)$. Instead of that, we prefer to determine the ratio Q of predators (that is, non-basal species) with adaptive foraging, and to have control over which species will have this character. Our hypothesis is that the larger the set of preys of a predator is, the more capable of adaption it is. (In the extreme case of a predator feeding only on one prey, adaptation is meaningless.) So, in our method, it will be important to order the predators in terms of the number of preys they have, and assign adaptation to the fixed fraction Q having more preys. A simple rank correlation reveals that the strong hierarchy of the food web furnished by the cascade model gives us the desired order. On the contrary, we don't find this correlation for the niche model: predators with high niche value can have few preys and vice-versa. The fractions considered are $Q = 0, 0.25, 0.5, 0.75$, and 1. The adaption rate g_i is taken 1 for all adaptive predators, and 0 otherwise.

The values of the parameters f_{ij} that capture the foraging efficiency of i feeding on j depend on the trophic class of both species, classified as top (T), intermediate (I) and basal (B). We then consider four possible values f_{TB}, f_{IB}, f_{TI} and f_{II} that are assigned to every f_{ij} depending on the classification of i and j (see Figure 1 and Table 1.) The parameters l_i are also dependent on this classification, with values l_B, l_I and l_T.

We fix the total number of species $N = 20$. We then let the system given by equations (1) and (2) to evolute for a large period (until $t = 2000$), when we remove species and links for which x_i or a_{kj} fall below 10^{-10}. If specie i is removed, so are all links emanating from/arraving to i (all a_{ij} and a_{ji}.)

For the numerical integration, we use the Runge-Kutta 4.5 method. Simulations are performed with MAPLE 10.

Results

We call *persistence* of a food web the rate of persistent species, and its *(directed) connectance* is $C = L/N^2$, where L stands for the number of links, and N for the number of species. In [20], the author studied the influence of foraging adaptation on the persistence of food webs that were generated with the cascade model. By using the DAM, he proved that the presence of adaptive predators can stabilize complex food webs, that is, there is a positive correlation between connectance and stability of the food web. These results were argued in [5], where substituting the cascade model by the niche model in order to generate the initial food web structure, the authors show that this correlation becomes negative. These works follow the classical complexity-stability debate ([24],[25] and [?]). However, in [24] and [25], the author studied the influence of the complexity (in terms of the connectance and the total number of species) on the stability of an equilibrium point in a static food web. The approach in [20],[21] and [5] is different: changes in the initial connectance of the food web imply changes in its persistence when the food web is close to an (hypothetic) equilibrium point. In other words, in the case of May, the connectance considered is the one at the equilibrium (which equals the initial because the food web structure is static), while [20],[21] and [5] uses the initial connectance in their analysis. In the sequel, we will differentiate if the connectance is initial or final.

Our starting viewpoint was to study the topology of the final ($t = 2000$) food web under the presence of foraging adaptation. So, initially, we took the set of parameters given in [5] (but fixing $g_i = 1$, see column A in Table 1) and proceed to do simulations using the DAM with the niche and the cascade model. For each selection of model (cascade or niche), connectance ($C = 0.19$ and $C = 0.4$) and fraction of adaptive predators ($Q = 0, 0.25, 0.5, 0.75, 1$), we generate 100 food webs.

We found the same type of results than [20] and [5]: with the assumption of the cascade model, correlation (initial) connectance-persistence is positive in presence of foraging adaption, and negative with the niche model. Our persistence indices are lower than the ones in [5], but we verify that this is due to the fact that their species are killed if its density is smaller than 10^{-30}, a (too) high permissive value in our opinion. We also saw the effect explained in [20]: for a given initial connectance, the connectance of the final food web decreases with the increase of the fraction of adaptive predators (Q), and this is also true for the niche model. Now, a careful inspection shows that the decreasing rate of the connectance are quite different for both models. Moreover, in the case of the cascade model the effect of foraging adaptation on the final connectance is the inverse of the mentioned (initial) connectance-persistence correlation: adaptive predators facilitate decreasing final connectance.

We also computed the proportion of adaptive and non-adaptive species that persist with respect to the initial adaptive-non-adaptive configuration, and both model share similar qualitative behaviors. For the cascade model, the proportion of adaptive species that survive is located around 70%, independently of the rate of adaptive predators in the food web, and the proportion of surviving non-adaptive species increase with increasing Q. For the niche model, we found that the rate of surviving adaptive species is not constant and decreases slightly with increasing Q.

We also analyzed the existence of trophic levels in the resulting food web in the following way: for each non-basal species, we computed the length of the shortest path in the resulting graph that links it with basal species, and we take the maximum value over all non-basal species to obtain the *trophic height* H of the food web, that is,

$$H = \max_{i \notin B} d(i, B),$$

Table 2: The module final connectance and persistence in terms of the set of parameters.

Set $A/B/C$	Connectance	Persistence (%)
$Q = 0$	0.17/0.5/1	50/75/100
$Q = 1$	0.33/0.5/1	75/100/100

where B is the set of basal species, and $d(i, B)$ is the distance from i to the set B. A height $H = h$ implies the existence of $h + 1$ trophic levels. For the choice A of parameters, the rich initial food web (averaging $H = 2$) collapses in a poor 2-trophic ($H = 1$) food web structure where all the persistent species are basals or feed on basal species directly. That occurs with both models, for all values of C and Q, and for nearly all generated food webs. A mathematical explicit analysis shows that with this choice of parameters, a simple tri-trophic food chain will not persist because of the top predator extinction. It seems that the species are too homogeneous and that homogeneity does not support complex structures.

We thought that other choice of parameters will allow to build more complex networks with three trophic levels, also enhancing persistence and final connectance. With this idea in mind, the test for our choice of parameters will be the stability of the trophic module given in Figure 1 (B), and the tuned parameters are f_{ij} and l_i. This module contains different types of predation (omnivory, intraguild predation) and the three trophic classes (T, I and B). The question is: can the stability and complexity of a food web be derived from the stability of a simple trophic module? That is a classical question posed in the 70's (see [25].)

After some tests, we started simulations with the set of parameters B and C of Table 1, and at that time we only consider the niche model. Table 2 summarizes the effect of each set on the module structure. Observe that from A to C, persistence and connectance of the trophic module increase for both values of Q. For $Q = 1$, sets B and C have the same persistence, but the connectance of C is twice the B one.

All the descriptors are quantitatively improved. With B and C sets we upgrade the trophic levels of the resulting food web reaching values between 2 and 3, observable in real food webs ([32]). The complexity of the final food web is clearly enlarged by the gain in link density (L/N) and number of links (L).

All the results will be included in two different papers and submitted for publication.

Discussion and future research

We have analyzed the final structure of a non-static food web whose dynamics is driven by the DAM in relation with the foraging adaptation. The results on the influence of adaptive predators are qualitatively different depending on the initial food web configuration and they must be explained in terms of the different topologies generated by the niche and the cascade models. Contrary to the strict trophic hierarchy of the cascade model, the niche configuration allows cannibalism and mutual predation giving rise to cycles or loops in the subjacent graph, a fact that is likely to contribute bringing light to these discrepancies. Mixing loops and foraging adaptation seems to derive in a decrease of the persistence in augmenting the (initial) connectance, while in the case of loops absence, more (initial) connectance equals more persistence. One possible explanation is that without loops, increasing connectance is like giving to adaptive predators the possibility to chose its food diet among a larger set of preys. If cycles are permitted, big connectances imply a lot of indirect effects on the population densities of species that can destabilize the food web if foraging adaption is considered.

This study must be extended considering other initial configurations as, for example, the one given by the nested hierarchy model. A mathematical analysis that could explain the influence of the initial topology in the resulting food web structure should be done.

Our contribution outlines that stability of the food web depends on stability of its sub-webs, and that complexity in food webs can be explained trough complexity of such sub-webs.

As can be derived from our results, an interesting fact appears: although adaptive predators facilitates persistence for a fixed initial connectance, big rates of foraging adaptation ($Q \geq 0,75$) diminishes the complexity and the richness of the food web, measured in terms of link density and trophic levels.

Finally, in [15] the authors argue that the usage of a type II Holling functional response do not permit to generate food webs with high trophic levels. In view of that, they use a ratio-dependent functional response to succeed, and they adopt the IAM. Since we have obtained good results with our model in that sense, a subsequent question is what are the differences in considering a ratio-dependence function response. It must be said that the IAM is a limiting case of the DAM. In particular, increasing unboundedly the adaption rate of species i (g_i) in equation (2) means that the species is capable of instant adaption, and that is the hypothesis in IAM. So, IAM supposes that the time needed for adaption is negligible with respect to the evolution time of the population densities, while the DAM equals behavioral and evolutionary time. The IAM is a consistent approach given in [13] to growing or evolving food webs, where new species are added to the food web by virtue of immigration or mutation, because the time scale is determined by evolutionary changes. One interesting question to be addressed is the influence on the evolutionary scale of a growing food web with the adoption of the DAM instead of the IAM. In particular cases, simulations with few species shows that from the dynamical perspective, there are no significant changes in substituting one by the other. But from the mathematical point of view, it is convenient to give theoretical results relating the DAM and the IAM, establishing similarities and differences in terms of the mentioned limiting convergence.

Bibliography

[1] K. Arii and L. Parrott. Emergence of non-random structure in local webs generated from randomly structured regional webs. Journal of Theoretical Biology **227** (2004), 327-333.

[2] J. Bascompte and P. Jordano. The structure of plant-animal mutualistic networks. In [28].

[3] J. Bascompte, P. Jordano and J.M. Olesen. Asymmetric coevolutionary networks facilitate biodiversity maintenance. Science **312** (2006), 431-433.

[4] J. Bascompte, C.J. Melián. Simple trophic modules for complex food webs. Ecology **86** (2005), 2868-2873.

[5] U. Brose, R.J. Williams, and N.D. Martinez. Comment on "Foraging adaptation and the relationship between food-web complexity and stability". Science **301** (2003), 918b.

[6] J. Camacho, R. Guimerà and L.A. Nunes Amaral. Analytical solution of a model for complex food webs. Physical Review E **65** (2002), 03090.

[7] ——— Robust patterns in food web structure. Physical Review Letters **88** (2002), 228102.

[8] M.F. Cattin, L.F. Bersier, C. Banasek-Richter, M. Baltensperger, and J.P. Gabriel. Phylogentic constraints and adaptation explain food-web strucuture. Nature **427** (2004), 835-839.

[9] X. Chen and J.E. Cohen. Global stability, local stability and permanence in model food webs. Journal of Theoretical Biology **212** (2001), 223-235.

[10] J.E. Cohen and F. Briand. Trophic links of community food webs. PNAS **81** (1984), 4105-4109.

[11] J.E. Cohen, F. Briand, and C.M. Newman. "Community food webs". Biomathematics vol. 20, Springer-Verlag, 1990.

[12] J.E. Cohen, T. Jonsson, S.R. Carpenter. Ecological community description using the food web, species abundance, and body size. PNAS **100** (2003), 1781-1786.

[13] B. Drossel, P.G. Higgs, and A.J. McKane. The influence of predator-prey population dynamics on the long-term evolution of food web structure. Journal of Theoretical Biology **208** (2001), 91-107.

[14] B. Drossel and A.J. McKane. Modelling food webs. In "Handbook of graphs and networks: From genome to internet", S. Bornhold and H.G. Schuster (Eds.), Wiley-VCH, 2003.

[15] B. Drossel, A.J. McKane, and C. Quince. The impact of non-linear functional responses on the log-term evolution of food web structure. Journal of Theoretical Biology **229** (2004), 539-548.

[16] J.A. Dunne, R.J. Willimans ad N.D. Martinez. Food-web structure and network theory: The role of connectance and size. PNAS **99** (2002), 12917-12922.

[17] J.A. Dunne. The network structure of food webs. Chaper 2 in [28].

[18] H.C. Ito and T. Ikegami. Food-web formation with recursive evolutionary branching. Journal of Theoretical Biology **238** (2006), 1-10.

[19] V.A.A. Jansen and G.D. Kokkoris. Complexity and stability revisited. Ecology Letters **6** (2003), 498-502.

[20] M. Kondoh. Foraging adaptation and the relationship between food-web complexity and stability. Sience **299** (2003), 1388-1391.

[21] ———— Does foraging adaptation create the positive complexity-stability relationship in realistic food-web structure? Journal of Theoretical Biology **238** (2006), 646-651.

[22] V. Krivan and O.J. Schmitz. Adaptive foraging and flexible food web topology. Evolutionary Ecology Research **5** (2003), 623-652.

[23] N.D. Martinez, R.J. Williams and J.A. Dune. Diversity, complexity, and persistence in large model ecosystems. Chapter 6 in [28].

[24] R.M. May. Will a large complex system be stable? Nature **238** (1972), 413-414

[25] R.M. May. "Stability and complexity in model ecosystems". Princeton University Press, 1974.

[26] C. J. Melián, J. Bascompte, and P. Jordano. Spatial structure and dynamics in a marine food web. In "Aquatic Food Webs", A. Belgrano et al. editors. Oxford University Press, 2005.

[27] S.A. Navarrete and E.L. Berlow. Variable interaction strengths stabilize marine community patterns. Ecology Letters **9** (2006), 526-536.

[28] M. Pascual and J.A. Dunne (Eds). "Ecological Networks: Linking structure to dynamics in food webs", Oxford University Press, New York, 2006

[29] P.C. de Ruiter, V. Wolters and J.C. Moore (eds.). "Dynamic food webs: Multispecies assemblages, ecosystem development and environmental change", Academic Press, 2005.

[30] D.B. Stouffer, J. Camacho, R. Guimerà, C.A. Ng, and L.A. Nunes Amaral. Quantitative patterns in the structure of model and empirical food webs. Ecology **86** (2005), 1301-1311.

[31] G. Sugihara, L.F. Bersier, T.R.E. Southwood, S.L. Pimm, and R.M. May. Predicted correponce between species abundances and dendrograms of niche similarities. PNAS **100** (2003), 5246-5251.

[32] R.J. Williams, E.L. Berlow, J.A. Dunne, A.L. Barabási, and N.D. Martinez. Two degrees of separation in complex food webs. PNAS **99** (2002), 12913-12916.

[33] R.J. Williams and N.D. Martinez. Simple rules yield complex food webs. Nature **404** (2000), 180-183.

Economic Networks

Data Acquisition on Network Formation in Economics

Roland Amann[1] and Thomas Gall[2]
[1] University of Konstanz, Fach D136, 78457 Konstanz, Germany and Thomas Gall
[2] University of Bonn, Economic Theory II
Lennéstr. 37, 53113 Bonn, Germany

Abstract
This section very briefly reviews some empirical strategies for the analysis of economic networks and presents a data set compiled for the studying groups model.

Introduction

The fundamental difficulty in obtaining data on contact networks lies, of course, in retrieving the links between individuals. This can be rather tedious at times owing to data protection regulations, individual reluctance to reveal information on social interaction, and the need to collect exhaustive samples. One approach to elicit information on contact networks is by simply asking individuals about their social interaction, that is by means of interviews and questionnaires. One of the more prominent data sets compiled by this method is the Add Health data set ([5]). It consists of a sample of 80 high schools and 52 middle schools from the US, representative with respect to region of country, urbanicity, school size, school type and ethnicity. Most interestingly for the analysis of social network formation the data set contains both information on friendship networks in high school and a follow-up study of socio-economic outcome of the sampled cohort. This information can be used to infer effects of network characteristics on individual outcomes such as criminal activities and educational attainment (see [2] for instance). Also [4] employs this approach in his analysis of means individuals had successfully used in encountering jobs.

Another approach that has been applied for instance by [6] is to use data generated by communication among individuals. In their case this is frequency of email contacts. The underlying assumption is that close social contacts require maintenance by sufficiently frequent communication.[1] Communication among individuals in the sample consisting of students of Dartmouth College, USA, is apparently carried out to a large extent by email, so that frequency of email interaction can be used as a proxy for link intensity. As a result in this study geographical distance between students' dormitory rooms is related inversely to the degree of friendship.

---

[1]Incidentally, this is exactly the classification used by [8] to distinguish between strong and weak ties.

Static versus Evolving Networks

This already suggests that compiling static data on individuals' contact networks can be relatively unassuming when covering students at school or university, for instance. Existing social infrastructure and geographical proximity can be exploited by making use of class meetings or accessing administrative data. The chief difficulty in gathering data on economic networks resides in collecting dynamic data on contact networks once individuals change their social environment. This has usually to be undertaken by means of rather cost-intensive follow-up interviews. A potential solution to this problem could lie in the use of data obtained from online contact networks, such as e.g. LinkedIn, provided regulatory obstacles can be overcome and companies can be persuaded to part with their data.

Measuring Studying Group Formation

Since our model ([1]), presented in section , is concerned with the formation of social networks used for human capital production rather than general friendship networks, the empirical analysis focuses on studying groups. Yet reliable data sets on endogenous formation of studying group do not exist to best of our knowledge. Therefore, we have compiled a unique working sample from six German universities in order to shed empirical light on the topic of assortative group formation. The participating universities are Bonn, Gießen, Frankfurt, Konstanz, Mainz and Mannheim. The collection of the data has proceeded in two stages. In class students were first asked to answer an individual questionnaire, and then to indicate links to other students they shared a studying group with. In order to be able to compare the cognitive skills across universities we have restricted our sample to students who major are in economics. In addition, we asked predominantly students who have been enrolled for more than two years. We believe that for these senior students the commitment of belonging to a certain studying groups is less biased by the lack of asymmetric information about their peers. Moreover, it allows us to control for the overall grade after the completion of the second year. We collected the data towards the end of the term before exams, thus increasing the likelihood of observing stable matchings and raising stakes.

The data set contains rich information about all relevant variables for the purpose of this study. First, we have several information about students cognitive skills. Next to the high school grade we asked the students about their grades in different examinations, the grade in the Vordiplom (intermediary examination) and the expected final grade. Secondly, the questionnaire contains many questions about students' preferences and social skills. For instance, we ask the students about their leisure time activities in order to elicit preferences for social versus non-social activities. Next to the information about individuals, the data set matches all information about individuals with their corresponding studying group. On average a group consists of almost three members and exists more than two semesters. Off campus group activities are taking place often for most of the groups. We have full information about almost 80 percent of all groups.

Participation

We are in particular interested in the effect of cognitive skills and social preference on the group participation. A first glance at the data (Figure) suggests an inverted u-shaped relationship between cognitive ability and participation. This is consistent with our model when human capital aggregation is important in studying groups and congestion stronger for high ability students.

In the empirical analysis, instead of assuming a linear trend in these two dimensions of students' characteristics we try to keep the model as general as possible and estimate semiparametrically the participation. We compute Probit estimates of the individual participation rate in studying groups

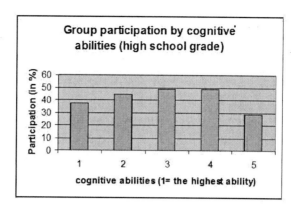

Figure 1: Participation rates depending on ability score

both using a linear specification and a non-linear specification introducing dummy variables for each possible value of the type. The estimates show that Germans and students who have taken a year of civil service are more likely a member of a studying group, as are females and active members of some club. Surprisingly, students being relatively new at university join a studying group with a higher probability. This might be explained by having the chance to meet people through a social learning environment emphasizing the social function of learning groups.

The parameters of interest are the effect of social and cognitive skills on the participation in studying groups. Indeed, students with better past grades are less likely to be in a group. On the other dimension students, who state to like meeting friends for cooking or board games – a proxy for preference for social interaction, are more likely to be in a studying group. That is, participation choice in studying groups appears to be governed by expected benefits incurred on both dimensions, social interaction and knowledge spillovers. Indeed, students are more likely to state that they join a group to be tutor the higher their ability and conversely for the role as a recipient of tutoring. Interestingly, the higher a student's score on the social dimension the likelier becomes the statement to be in for being tutored although ability and social dimensions are not correlated in the data.

That is, the main assumptions that the studying groups model builds on appear to be satisfied in our sample. Individuals obtain benefits from the interaction in studying groups on more than one dimension and participation choice is rational and based on expected gains on these dimensions.

Group Composition

The first point of interest is whether there emerges any systematic pattern, for instance whether highly able students tend primarily match among themselves. In Figure frequency of studying groups by degree of homogeneity, that is in-group variance of ability score, is shown for a split of the sample into three different ability groups. The first important observation is that there are heterogeneous studying groups, an indicator that some compensation is possible between low and high ability agents. Moreover while the best students tend to keep by themselves, there appears to be mixing between students of intermediate and low cognitive ability.

The observations in Figure are confirmed when we compute Tobit regressions of mean ability

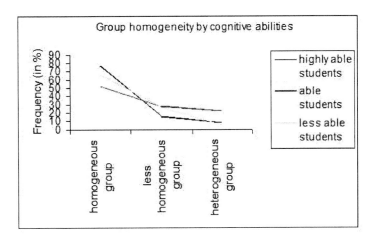

Figure 2: Frequency of different degrees of group homogeneity

of an individual's peers using a non-linear specification of individual attributes. As controls we use socio-economic characteristics and a measure of proximity of tastes computed from the degree of in-group coincidence of answers to lifestyle and taste questions. Indeed, peer ability appears to be higher for low ability students than for intermediate students indicating the presence of mixing. Moreover, peer ability is found to depend negatively on own score on the social dimension consistent with our theoretical model.

Bibliography

[1] Amann, R. and T. Gall (2006): 'How (not) to Choose Peers in Studying Groups'. *FEEM Working Paper* No. 79.2006.

[2] Calvó-Armengol, A., E. Patacchini and Y. Zenou (2005): 'Peer Effects and Social Networks in Education and Crime'. *Mimeo Universitat Autononoma de Barcelona.*

[3] Granovetter, M. (1973): 'The Strength of Weak Ties'. *American Journal of Sociology* 78, 1360-1380.

[4] Granovetter, M. (1995): *Getting a Job: A Study of Contacts and Careers*, 2nd edition, University of Chicago Press, Chicago.

[5] Harris, K.M., F. Florey, J. Tabor, P.S. Bearman, J. Jones and J.R. Udry (2003): 'The National Longitudinal Study of Adolescent Health: Research Design'. URL: http://www.cpc.unc.edu/projects/addhealth/design.

[6] Marmaros, D., and B. Sacerdote (2004): 'How do Friendships Form?'. *Mimeo Dartmouth College.*

Solution Concepts in Economic Network Formation Models

Thomas Gall and Benny Moldovanu
University of Bonn, Economic Theory II
Lennéstr. 37, 53113 Bonn, Germany

Abstract
This section provides a brief overview of solution concepts prevalent in network formation models in economics.

Introduction

Important aspects of the economic allocation are often crucially determined by the allocation of social ties between economic agents, that is the formation of social networks. Nodes of an economic network are typically the set of individuals who decide whether to link to other individuals based on some cost-benefit-calculation. Links between agents may represent for instance mutual trade, information exchange, or insurance. For instance, in many economies an individual's success in encountering suitable employment depends largely on the social ties this individual has.

Studies pioneered by [17] find that individuals frequently encounter new jobs on the base of personal relationships. An individual with a greater number of loose acquaintances stands a better chance of encountering new employment [8]. Other examples include trading networks where trade opportunities are open only to network members, or implicit risk-sharing agreements in social groups (for a recent survey on social network formation see [10] and [11]). Therefore economic efficiency of an allocation cannot be evaluated independently of the formation of such social networks.

Graph Theory - Network Topologies

Network formation and evolution in economics is typically based on rational choice of utility maximizing individuals. Hence economic network topologies often result from individual behavior, that is the choice of links by economic agents. This means nodes in a network represent economic agents and links can be interpreted as passing on information, cooperating on a project or engaging in a trade relationship, depending on the economic situation of interest. The concept of economic agent encompasses any smallest decision-making entity, usually individuals.[2] In order to determine

---

[2]See e.g. [7] for an example where nodes represent firms and links cost decreasing collaboration.

outcomes – be it in terms of the economic allocation or the network actually formed – a solution concept for network formation models is needed. Ideally it should be able to predict unique outcomes depending on the initial state of the model for all initial states of the model. In the following we provide a brief review of some concepts that have proven pervasive in the economic literature on network formation so far.

Example: To choose a simple example (the connections model from [5]), suppose for instance there is a set of agents $N = 1, 2, ..., n$. Links between agents represent the flow of information. Assume links are undirected, that is a link between agent i and j enables information flow in both directions and requires mutual consent. Denote a link between agents i and j by ij. A link ij requires fixed maintenance cost c that has to be born by both agents. Links between agents are representable by a non-directed graph. Let g denote the graph, $g = (N, L)$ where L denotes a set of links between agents. Let $g^C(N)$ denote the complete graph over N, l^C the corresponding set of links and $\mathcal{G}(N) = \{(N, L)|L \subseteq L^C\}$ the set of all feasible graphs over N. A graph g is completely connected if for any tuple (i, j) there exists a sequence of links $\{ik_1, k_1k_2, ..., k_nj\}$, that is a path connecting i and j.

Information is an excludable, non-rival good that induces utility. This can be thought of as a model of knowledge spillovers, where information acquisition increases productivity. Quality of information transmission deteriorates as it is passed along nodes. Therefore path length $d(i, j, g)$ between two nodes i and j in graph g, that is the number of links in the shortest path connecting i and j, matters. Set $d(i, j) = \infty$ if i are not connected. Then agent i's utility in graph g may be represented by $u_i(g) = \sum_{j \neq i \in N} \delta^{d(i,j,g)} - \sum_{j:ij \in g} c$, where $\delta \in (0, 1)$ represents imperfect information transmission. That is, an agent's utility depends both on the number of direct links and on the number of indirect links.

Stochastic processes defined on network topologies

Given an economic network agents make use of the links they hold. As a consequence agents' attributes may change, for instance information sets, investment goods such as (human) capital, or wealth. That is, there is a state space attached to the network topology and state changes of the economy depend on the network topology. If using links entails uncertainty, for instance because of stochastic production or the use of mixed strategies in games between linked agents, the state of the economy may follow a stochastic process. Hence, at the time of network formation agents anticipate that the properties of the stochastic process will depend on the network formed. Note that this may introduce a non-trivial feedback into the problem of network formation.

Attaching state spaces to graphs

Usually, the relevant state space for economic models is based on individuals' attributes, such as wealth, employment or health status. That is, the state space can be constructed based on the nodes. In the simplest case, for static economic network formation models, the state space attached to the network is relevant only in terms of expectations formed ex ante by agents when deciding on links. This gives expected continuation payoffs depending on own choice and potentially all remaining agents' choices. Let us review possible solution concepts for the static setting in the following.

Nash Equilibrium

The most immediate way to determine outcomes of network formation is to employ a game-theoretic model and use the Nash equilibrium. That is, agents choose actions according to some preference ranking over outcomes taking into account that other agents do just the same. The space of actions could e.g. be given by a set of agents, indicating whether to maintain a link to another agent or not (see [16], chapter 9, and [6]). Then it is possible to determine for each agent the set of actions that is a best response (i.e. that yields the preferred outcome) for any given tuple of the remaining agents' actions. A Nash equilibrium is a fixed point of agents' best response correspondences. This concept has one potential drawback, however, in that the set of networks resulting from a Nash equilibrium is quite large, and it always allows for the empty network in static games whenever mutual consent is required to establish a link. This is undesirable in an outcome from an economic point of view since one could expect agents to find a way to overcome such coordination failures if the potential gain from link formation was sufficiently great.

Example: In case of the simple information transmission example, the corresponding non-cooperative linking game is given as follows. Agent i's action is denoted by a_i and i's action space is given by $a_i \subseteq N \setminus \{i\}$. A link between agent i and j emerges if both agents choose to link, that is $ij \in g$ if both $\{i\} \in a_j$ and $\{j\} \in a_i$. A graph g is the outcome of a Nash equilibrium if given g, for all $i \in N$ it holds that

$$\sum_{j \in a_i : i \in a_j} \left(\delta^{d(i,j,g)} - c \right) \geq \sum_{j \in a_i' : i \in a_j} \left(\delta^{d(i,j,g)} - c \right) \forall \, a' \neq a.$$

A graph g is said to be the outcome of a strict Nash equilibrium if all of the inequalities above are strict. Note that $u_i(a_i, a_{-i}) \geq u_i(a', a_{-i})$ holds for actions a_i and a_i' that differ only in that for all $j \in a_i' \setminus a_i$ $i \notin a_j$. It follows immediately that for sufficiently small c given δ all possible graphs can be supported as an Nash equilibrium outcomes. Note that this is not true for the strict Nash equilibrium, since for instance the empty graph cannot supported as an equilibrium any more.

Pairwise Stability

[5] propose a stricter solution concept, pairwise stability. This is a static concept as well. Unlike the Nash equilibrium this notion of stability does not formally require agents to choose actions but is rather a property of network formation outcomes. Any particular network has the property of pairwise stability if there do not exist (i) two agents not linked in the particular network both preferring to be linked ceteris paribus, nor (ii) an agent preferring to sever any one link present in the particular network ceteris paribus. Whereas for networks resulting from a Nash equilibrium only deviations by any single agent must be unprofitable, for pairwise stability also joint deviations by any two agents establishing a new link must be unprofitable. Under pairwise stability the empty network is only an outcome if the cost of link formation outweighs the benefits for at least one agent for every potential link, reflecting economic intuition. For instance, [3, 3] use this concept on the job search through social contacts problem outlined above. However, existence of a pairwise stable network in any particular network formation model is not guaranteed and if a pairwise stable network exists, by no means does it need to be unique. This is particularly unpleasant when the setting of interest is dynamic.

Example: Let us return to the information transmission example. Given a graph g denote the operations of adding a link ij to g by $g + ij$ and subtracting a link ij from g by $g - ij$. A graph g is said to be pairwise stable if given g, for all $i \in N$ it holds that (i) for $ij \in g$:

$$\sum_{k \neq i \in N} \delta^{d(i,k,g)} - \delta^{d(i,k,g-ij)} \geq c \text{ and } \sum_{k \neq j \in N} \delta^{d(j,k,g)} - \delta^{d(j,k,g-ij)} \geq c.$$

and (ii) for all $ij \notin g$:

$$\sum_{k \neq i \in N} \delta^{d(i,k,g)} - \delta^{d(i,k,g+ij)} > c \Rightarrow \sum_{k \neq j \in N} \delta^{d(j,k,g)} - \delta^{d(j,k,g+ij)} < c.$$

For sufficiently low cost c given δ the set of pairwise stable allocations is indeed a singleton, the complete graph. Intuitively, for low enough cost the gain in transmission quality makes forming a link between any two individuals profitable, and since the concept allows for pairwise deviations these gains have to be realized in equilibrium. This is in stark contrast to the set of Nash outcomes.

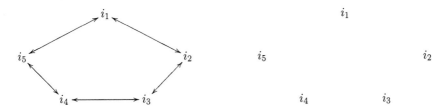

Figure 1: Stable network architectures for intermediate cost

For intermediate cost, however, profitable deviations that might require rewiring more than one link cannot be captured by this concept, allowing for larger equilibrium outcome sets. It is easy to see that for this reason pairwise stability does not imply Pareto efficiency.[3] Suppose for the sake of concreteness that $n = 5$ and $\delta < c < \delta + \delta^2(1 - \delta - \delta^2)$, implying that $\delta < (\sqrt{5} - 1)/2$. This means a link is only valuable if it provides access to at least a second node. It is easy to see that all pairwise stable networks are characterized by the property that either each agent has no links or each agent has two links as depicted in Figure 1. This coincides with the set of Nash outcomes. Note that the efficient network architecture, the star, is not in the equilibrium outcome set.

Change of state

Although the dynamics of the economy given a network is only relevant to network formation as far as it affects continuation payoffs the reverse does not need to hold. The nature of the network formed in equilibrium may have considerable impact on the subsequent dynamics of the economy. For instance, an empty graph may lead to stagnation for lack of trade whereas a completely connected graph may allow the realization of gains from trade and thus promote growth.

Discrete dynamical systems on networks

As the state of the economy, e.g. the matrix of agents' employment status, changes some agents may wish to adjust their portfolio links. Incorporating this into a model of economic network formation leads from static networks to evolving networks. For instance, when links represent engagement in trading relations individuals may find it profitable to condition the continuation of the relationship on past behavior. In other words, link adjustment may be used to punish defecting individuals (see e.g. [9] who describes a long-distance trading network where the threat of withdrawing a link serves to discipline traders).

Linking decisions are taken to maximize individual expected future payoff, that is agents labor to secure favorable individual transition probabilities. In case of evolving networks this means individual choice takes into account future link adjustments of other agents. A concept based on discrete time that has been proposed in the literature to provide solutions to economic network formation models taking into account possible evolution of the network is farsighted stability.

---

[3]A concept that postulates stability with regard to coalitional deviations under perfectly transferable utility is the Myerson value ([15]).

Farsighted Stability

Nash equilibrium, pairwise stability and stability with respect to myopic best response dynamics require an outcome network to be robust against some deviations only. First, only deviations by a small number of agents (i.e. one or two) are considered. Second, indirect consequences of deviations are not considered. For instance, if forming a new link is profitable for two agents this may generate a network that makes it profitable for a third agent to sever one existing link. This reasoning lies at the heart of the concept of farsighted consistent networks (see [14], [18]).[4] The notion postulates that a network in the outcome set of a model has to be stable with respect to profitable sequences of deviations (as opposed to sequences of profitable deviations). Indeed, a set of farsighted consistent networks generally exists. This set may, however, consist of multiple networks forming a cycle of feasible deviations.

Example: Let us turn back to the example of information transmission. Focus on the intermediate cost setting with $n = 5$ and $\delta < c < \delta + \delta^2(1 - \delta - \delta^2)$. Farsighted stability requires an outcome to be stable with respect to finite sequences of profitable coalitional deviations. Recall that the set of pairwise stable networks was given by the empty graph and all graph such that each node has exactly two links. Let us now check whether the empty graph is farsightedly stable. To show that it is indeed not, take a coalition $S = \{1, 2, 3, 4, 5\}$, that is the grand coalition, and check for individually profitable deviations. Propose the following deviation: establish links 12, 23, ..., 51. Simple computation shows this to yield higher utility for each agent in the deviating coalition. Feasibility trivially holds since S is the grand coalition.

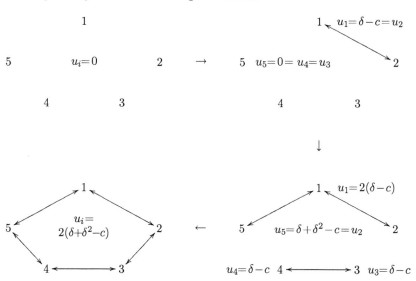

Figure 2: A sequence of coalitional deviations under farsighted stability

To further illustrate the concept Figure 2 depicts in clockwise order a randomly picked sequence of coalitional deviations that ends at $\{12, 23, ..., 51\}$ giving all agents that had to adjust their links higher utility than at the time of readjustment (i.e. the sequence is a farsighted domination path).

---

[4]See also a similar approach taken by [5]. In their concept a single agent readjusts his link portfolio per period fully taking into account all possible consequences in future periods.

Utilities of agents are denoted as subscripts. The sequence indeed terminates at $\{12, 23, ..., 51\}$ since there is no coalitional deviation with another network architecture that assigns strictly positive utility to all agents and moving to another ring architecture gives raise to the same utility for all players. That is, any completely connected graph with two links per agent is farsightedly stable. Note that the efficient architecture, the star, is not stable under this concept either. A singleton, the hub, has a feasible and profitable (that is, profitable with respect to terminal graph of the sequence of deviations) deviation to the empty network, and then the grand coalition has a profitable deviation as shown above.

Non-deterministic dynamics on networks

Due to rational choice of individuals in link formation an economic model of network formation typically entails deterministic dynamics of the graph. There exists some opportunity for non-deterministic network evolution when the timing of agents' decisions is random or agents are allowed to use mixed strategies when indifferent. For instance, the contribution of [2] has both these features. Steady state networks are characterized by strict individual preference, however.

Dynamic Stability

Another approach to solve models of network formation is to postulate dynamic stability of the outcome using a discrete dynamics. This may be achieved, for instance, by starting with a given network and in every following period altering links according to pre-specified rules. An example for this approach is [2] (see also [4]). They postulate that a stable network is a steady state of a dynamic system governed by the following rule. Every period a set of agents is chosen randomly and has the opportunity to severe existing or establish new links unilaterally without taking into account the behavior of other agents. Since agents choose actions that lead to their most preferred outcome but do not behave strategically this has been called myopic best response dynamics. [2] find that networks that are strictly preferred by all agents to all other networks are absorbing states for this dynamic systems and that indeed for sufficiently simple formulations of the model the the steady state network has this property almost surely.

Conclusion

To conclude, we observe that all stability concepts mentioned above suffer to some extent from non-existence and/or multiplicity problems. The concept related most closely to stability of networks in other disciplines is the notion of dynamic stability, in particular when agents do not behave as infinite horizon utility maximizer but rather exhibit myopia. This solution concept can be applied to problems of various degrees of complexity including models of stochastic dynamic network formation where both the network and states of the nodes may evolve – a rather common feature of economic applications such as job search through social contacts. Thus the use of an adequate version of dynamic stability appears to be a promising direction for research when pursuing a unified approach to modeling networks.

Bibliography

[1] Bala, V. and S. Goyal (2000): 'A Noncooperative Model of Network Formation'. *Econometrica* 68(5), 1181-1229.

[2] Calvó-Armengol, A. (2004): 'Job Contact Networks'. *Journal of Economic Theory* 114(1), 191-206.

[3] Calvó-Armengol, A. and M.O. Jackson (2006): 'Networks in Labor Markets: Wage and Unemployment Dynamics and Inequality'. *Journal of Economic Theory (forthcoming)*.

[4] Calvó-Armengol, A. and Y. Zenou (2005): 'Job Matching, Social Network and Word-of-mouth Communication'. *Journal of Urban Economics* 57(3), 500-522.

[5] Dutta, B., S. Ghosal and D. Ray (2005): 'Farsighted Network Formation'. *Journal of Economic Theory* 122(2), 143-164.

[6] Dutta, B., A. van den Nouweland and S. Tijs (1998): 'Link Formation in Cooperative Situations'. *Interbational Journal of Game Theory* 27, 245-256.

[7] Goyal, S. and S. Joshi (2003): 'Networks of Collaboration in Oligopoly'. *Games and Economic Behavior* 43(1), 57-85.

[8] Granovetter, M. (1973): 'The Strength of Weak Ties'. *American Journal of Sociology* 78, 1360-1380.

[9] Greif, A. (1993): 'Contract Enforceability and Economic Institutions in Early Trade: The Maghribi Traders' Coalition'. *American Economic Review* 83(3), 525-548.

[10] Jackson, M.O. (2005): 'The Economics of Social Networks'. *mimeo*, California Institute of Technology.

[11] Jackson, M.O. (2005): 'A Survey of Network Formation Models: Stability and Efficiency' in: *Group Formation in Economics*, eds. Demange, G. and M. Wooders, Cambridge University Press, Cambridge, 11-57.

[12] Jackson, M.O. and A. Watts (2002): 'The Evolution of Social and Economic Networks'. *Journal of Economic Theory* 106(1), 265-295.

[13] Jackson, M.O. and A. Wolinsky (1996): 'A Strategic Model of Social and Economic Networks'. *Journal of Economic Theory* 71(1), 44-74.

[14] Kamat, S., F.H. Page, Jr. and M.H. Wooders (2005): 'Networks and Farsighted Stability'. *Journal of Economic Theory* 120(2), 257-269.

[15] Myerson, R. (1977): 'Graphs and Cooperation in Games'. *Math. Operations Research* 2, 225-229.

[16] Myerson, R. (1991): *Game Theory: Analysis of Conflict*. Harvard University Press, Cambridge, MA.

[17] Myers, C.A. and G.P. Shultz (1951): *The Dynamics of a Labor Market*. Prentice-Hall, New York.

[18] Page, Jr., F.H. and M.H. Wooders (2005): 'Strategic Basins of Attraction, the Farsighted Core, and Network Formation Games'. *FEEM Working Paper Series* No. 36-2005.

Two Applications of Social Networks in Economics

Thomas Gall
University of Bonn, Economic Theory II
Lennéstr. 37, 53113 Bonn, Germany

Abstract

This overview presents two applications of social network formation in economic contexts. The studying group model analyzes social group formation with multidimensional agents, finding that stable matching patterns are not monotone. The job opportunity model presents a unified framework for analysis of the dynamic formation of social networks used for insurance and information transmission under uncertainty.

Introduction

In the following we present two applications of the formation of social networks in economic contexts. As usual for economic network nodes represent individuals. Links are undirected and enable cooperation between two nodes. This determines expected continuation payoffs to the nodes and thus individual link decisions. That is, the network topology is endogenous and follows from individual utility maximization. The first example is concerned with properties of a static equilibrium network. The second example admits possibly non-deterministic dynamics concerning both changes of the state and the network topology.

Consider first the endogenous formation of productive teams, characterized by exclusive subnetworks. We argue that individuals' types may be multidimensional, so that an agent's relative weakness on one dimension may be compensated for by a relative strength on another dimension. We show that this intuition indeed holds true and matching is positive assortative only with respect to an aggregate of type dimensions. This finding yields testable predictions on the outcomes even when only part of the type space is observable. Using a unique data set on the formation of studying groups in German universities we provide a first empirical test of the model.

The second application we have in mind concerns the labor market. Individuals' success in job search is highly affected by the information they obtain through social contacts. The value of a particular contact to an individual and thus the individual's choice of link formation will then depend potentially not only on the entire contact network but also on the employment state of the whole agent set. A link to an employed agent who maintains many unemployed contacts may be less desirable than a link to an employed agent who maintains few links to employed agents. We aim to characterize the steady states of such a dynamic model of network formation.

Static Network Topologies: Studying Groups Model

This section presents a simple example of the studying groups model of [1]. In a static setting exclusive networks of agents form following individuals' link decisions based on utility maximization. That is, the model identifies equilibrium network topologies given continuation payoffs. Agents obtain utility in this model both instantaneously from social interaction and from expected future income – which possibly determines future social interactions as well. We call the former source of utility *consumption peer effects* and the latter *production peer effects*.

Model Setup

An economy consists of a continuum of agents. An agent is characterized by his ability to generate consumption and production peer effects, that is a tuple $(\gamma_i, \theta_i) \in I \equiv \{\gamma_L; \gamma_H\} \times \{\theta_L; \theta_H\}$, either of which may be either high or low. Let $9/8 < \gamma_L < \gamma_H < 2$ and $9/8 < \theta_L < \theta_H < 2$. Denote such a tuple by i. Let each realization $i \in I$ be endowed with equal Lebesgue measure. Call I the agent set. An agent i's preference ordering over exposure to consumption and production peer effects can be expressed by an additively separable function, say $c_i + h_i$, where c_i denotes consumption peer effects and h_i production peer effects agent i is exposed to.

Agents may form exclusive and complete social networks to benefit from peer effects. A social network of individuals in this model can be interpreted as a social group fully encompassing any direct utility benefits emerging from social interaction between agents. A social network N is the set of its members $N = \{(\gamma_1, \theta_1), ..., (\gamma_i, \theta_i), ..., (\gamma_n, \theta_n)\}$ or, equivalently, the set of its members' types. The size of network N is given by n. We assume that each member has the possibility to unilaterally withdraw from a network and all members of a network must consent to the entry of a new member. Within networks agents obtain utility according to the peer effect functions $c_i(N)$ and $h_i(N)$ possibly depending on the types of all members of network N.

Define agent i's valuation function as $v_i = c_i(N) + h_i(N)$ and assume a functional form of c_i and h_i. For the sake of simplicity we impose some symmetry and define v_i as follows.

$$v_i(N) = \left(\sum_{j \in N} \gamma_j \right)^{\frac{1}{n}} + \left(\sum_{j \in N} \theta_j \right)^{\frac{1}{n}}.$$

This means an agent's valuation for joining network N is the utility from being exposed to the peer effects in the network. If an agent remains solitary his valuation is given by $\underline{v}_i = v_i(\{i\}) \geq 9/8$. Moreover, there are congestion effects present entering the valuation function via the exponent. This may, for instance, be due to agents' limited span of attention. Note that all agents have the same preference ranking over network composition in this example. For a general approach allowing for differing tastes see [1].

Group formation occurs on a matching market under complete information. Agents decide on whom to match with and can commit to their decision. Network membership is strictly voluntarily. We are looking for a stable matching of agents. A partition of the agent space into finitely sized social networks is called stable if there does not exist a deviating social network such that all its members have strictly higher valuation than under the stable partition.

Pareto optimality has not much bite in this framework, as introducing an arbitrarily small degree of transferability into the model renders any matching equilibrium allocation Pareto optimal. Therefore we look at aggregate valuations of agents. In a world with perfectly transferable utility matching will be negative assortative on each dimension giving the maximum of aggregate valuations.

Properties of the stable allocation

Note first that $v_i(N)$ is strictly increasing in each attribute of the members of N. Hence, there cannot be a positive measure of network involving at least on agent of type (γ_H, θ_H) and at least one agent of any other type. Suppose the contrary. Then a positive measure of types (γ_H, θ_H) has a profitable deviation by forming homogenous networks of the same size. Hence, the highest types segregate.

A similar argument holds for the lowest types (γ_L, θ_L). Suppose there is a positive measure of networks involving at least on agent of type (γ_L, θ_L) and at least one agent of any other type. By monotonicity of $v_i(.)$ in types agents of other types have at least one profitable deviation by forming homogeneous networks of the same size. That is, the lowest types segregate as well.

Turn now to heterogeneous types. Since $v_i(N)$ is submodular in types for fixed network size n, there cannot be a positive measure of homogenous networks of both (γ_H, θ_L) and (γ_L, θ_H) with size greater than one. Suppose the contrary. Then there exists a profitable deviation by forming heterogenous networks of the same size containing at least one agent of each type (γ_H, θ_L) and (γ_L, θ_H). Hence, heterogenous types from heterogenous networks or stay solitary, depending on the severity of congestion.

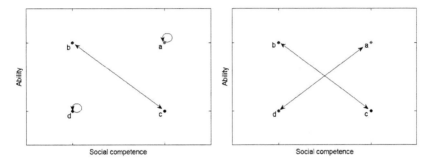

Figure 1: Outcome of decentralized network formation (left) and efficient network (right)

Given the assumption on parameters it can be checked easily that the optimal size of networks is indeed two for all type constellations and the stable allocation has measure $1/8$ of both $\{(\gamma_L, \theta_L), (\gamma_L, \theta_L)\}$ and $\{(\gamma_L, \theta_L), (\gamma_L, \theta_L)\}$ networks and measure $1/4$ of $\{(\gamma_L, \theta_H), (\gamma_H, \theta_L)\}$ networks. See Figure 1 (left) for an illustration of the network allocation formed in the example. The efficient network allocation which corresponds to a stable allocation under perfectly transferable utility is depicted to the right. The property that links tend to be formed along type level sets carries over to more general version of the model and delivers a testable prediction on the network topology.

Conclusion

In a decentralized social network formation model where agents' types are multidimensional and utility is non-transferable, a stable allocation of agents into social networks has segregation for extreme types and mixing for intermediate types. This is in marked contrast with the allocation under full utility transferability which has negative assortative matching on each dimension and the allocation given a one-dimensional type space which has positive assortative matching in types. Agents matching into heterogenous social networks are the ones to gain most from network participation since they experience absolute gains from trade. A preliminary analysis of a data set described in section ?? appears to be consistent with these findings.

The general setup can be found in [1]. They generalize the model to show that the result carries over whenever there are absolute gains from trade between some types of agents. Moreover, the setup can be extended to include effort investments on each dimension to allow for in-kind payments. Then the result carries over for sufficiently severe non-transferabilities of utility.

Evolving Networks with State Change: Job Opportunity Model

This section presents the framework for a dynamic analysis of the formation of social networks used for information transmission of vacancies. Individuals are the nodes of the network and a link between two nodes enables information flow on vacancies. Individuals can be employed or unemployed which gives the state space of the model. Note that the state changes according to transition rules that depend on the probability of being informed of a vacancy and thus on the network topology. Since agents have the opportunity to adjust their links and may use mixed strategies the network topology itself may follow a stochastic process.

Model Setup

Start with a set of agents, $N = \{1, ..., n\}$. Each agent is either employed or unemployed, that is $e_i \in \{0; 1\}$. Denote the vector of individual employment states by e. There is a network connecting the agents that is representable by a non-directed graph. Let G denote the graph, $g = (N, L)$ where the graph's vertices consist of the agent set N and a set of links between agents L. L is a set of *unordered* tuples of agents. There is a link between agents i and j if and only if $(i, j) \in L$. Denote by $\mathcal{N}^N(i)$ the set of direct neighbors of i, that is $\mathcal{N}^N(i) = \{j \neq i \in N | (i, j) \in L\}$. Let $n^N(i)$ denote its size. Likewise define by $\mathcal{N}^L(i)$ the set of links that has i as a starting or end point, $\mathcal{N}^L(i) = \{(i, j) \in L | j \in N\}$. Let $n^L(i)$ denote its size. Let $G^C(N)$ denote the complete graph over N, L^C the corresponding set of links and $\mathcal{G}(N) = \{(N, L) | L \subseteq L^C\}$ the set of all feasible graphs over N.

Agents have to choose who to link to and who not to. Denote an agent i's strategy by $s^i \in \{0; 1\}^n$. A link is formed between agents i and j by mutual consent, that is $(i, j) \in L$ if and only if $s_j^i s_i^j = 1$. Hence, G depends on the strategy profile. A link $(i, j) \in L$ requires maintenance cost $c > 0$ that has to be born by i and j each. Links transmit information on vacancies. With probability $\alpha \in (0; 1)$ an agent i loses his job, setting his employment state to $e_i = 0$. With probability $\beta \in (0; 1)$ each agent in N receives an exogenous job offer. If the agent is unemployed, he takes the job and his employment state changes to $e_i = 1$. If the agent is employed, that is if $e_i = 1$, the agent passes the information to agents directly linked to i, that is to $j \in \mathcal{N}^N(i)$. If more than one of i's direct links is unemployed, one of them is selected with equal probability and gets employed[5]. Denote the range of information transmission by $r \in \mathbb{N}$. r specifies how many times information of vacancy can passed on by employed agents. If $r = 1$ a vacancy can be passed on just once. If $r \geq 1$, a vacancy is passed on until an unemployed agent is in the set of direct neighbors, however, at most r times.[6]

Agents' preferences can be expressed using a utility function $u_i = e_i - n^L(i)c$. That is an agent's utility depends on the number of links and on the individual's employment state at the end of the period. Since e_i is random variable at the time of choice of s_i, agents form expectations

---

[5]This includes the possibility that an unemployed agent obtains a job offer from the market and receives another one through the contact network.

[6]This is very similar to the problem studied by [3].

on the realization of e_i which potentially depends on the graph G and on agents' states at the beginning of the period. That is, we define $Eu_i : \mathbb{R}^{n \times n} \times \mathbb{R}^n \mapsto \mathbb{R}$.

The timing within a period t_k, $k \in \mathbb{N}$ is as follows.

(i) Agents choose their strategies resulting in an updated graph G_{t_k}.

(ii) Nature draws job losses and vacancies yielding an updated employment state vector e_{t_k}.

This means there are two updates, one on the graph based on $G_{t_{k-1}}$ and e_{t-1}, and another one based on the updated graph G_t and e_{t-1}. We can write this as

$$(G_{t_k}, e_{t_k}) = (T_G(G_{t_{k-1}}, e_{t_{k-1}}), T_e(T_G(G_{t_{k-1}}, e_{t_{k-1}}), e_{t_{k-1}}),$$

where T_e denotes the employment operator, $T_e : \mathbb{R}^{n \times n} \times \mathbb{R}^n \mapsto \mathbb{R}^n$ and T_G the network operator $T_G : \mathbb{R}^{n \times n} \times \mathbb{R}^n \mapsto \mathbb{R}^{n \times n}$.

When giving structure to the network operator T_G, one could essentially take one of two possible views. First, agents' behavior may be assumed such that in every period t_k some equilibrium is reached, that is the strategy profile s_{t_k} determining the graph G_{t_k} has certain properties, for instance pairwise stability. Second, agents may be assumed to exhibit adaptive or myopic behavior, that is s_{t_k} giving rise to G_{t_k} will not – or only to a limited extent – take into account any expectations or reasoning might have on other agents' behavior.

Equilibrium updating

Start with the case where the network operator is determined by the assumption that agents' individual behavior is in equilibrium. It is then generated by an equilibrium property, such as pairwise stability [5] which is defined as follows.

Definition 1. *A network represented by G is pairwise stable given employment state e if and only if*

(i) *for all $(i, j) \in L$ $Eu_i(G, e) \geq Eu_i(G^-, e)$ and $Eu_j(G, e) \geq Eu_i(G^-, e)$ where $G^- = (N, L \setminus (i, j))$, and*

(ii) *for all $i, j \in N$ with $(i, j) \notin L$ it holds that $Eu_i(G^+, e) > Eu_i(G, e)$ implies $Eu_j(G^+, e) < Eu_j(G, e)$ where $G^+ = (N, L \cup (i, j))$.*

Under the assumption that agents' behavior gives rise to such an equilibrium in each period the network operator can be defined as

$$T_G(G_{t_{k-1}}, e_{t_{k-1}}) = \{G \in \mathcal{G}(N) | G \text{ pairwise stable under } e_{t_{k-1}}\}.$$

Note that for this case T_G does not depend on $G_{t_{k-1}}$ and we can write

$$(G_{t_k}, e_{t_k}) = (T_e(T_G(e_{t_{k-1}}), e_{t_{k-1}}), T_G(e_{t_{k_{k-1}}})).$$

Adjustment updating

On the other hand, individual behavior may be governed by adaptive ruled. In this case the network operator is generated by applying the chosen adaptive rule to each agent. One such rule is for instance given by a simple best response dynamics as in [2]. It is defined as follows. Select a tuple of agents $N^S \subseteq N \cup \emptyset$ with selection probability π for each $i \in N$.

Definition 2. *Under simple best response dynamics, for all $i \in N$ agent i's strategy is given by*

$$
s_i^{BG} = \begin{cases} \operatorname{argmax}_{s_i \in \{0;1\}^{n-1}} Eu_i(G_{t_{k-1}}(s_i, s_{-i_{t_{k-1}}})|e_{t_{k-1}}) & \text{for all } i \in N^S \\ \left(I\left((i,j) \in L_{t_{k-1}}\right)\right)_{j \in \mathcal{N}_{t_{k-1}}^N(i)} & \text{otherwise.} \end{cases}
$$

In the above definition $I(.)$ denotes an indicator function returning 1 when the argument is true and 0 otherwise. The probability π slows down the adjustment process. In case $|s_i^{BG}| > 1$ one element from s_i^{BG} is selected randomly with equal probability yielding a single-valued s_i^{BG}. Then T_G can be defined as

$$
T_G(G_{t_{k-1}}, e_{t_{k-1}}) = \{G \in \mathcal{G} | (i,j) \in L \Leftrightarrow s_{ij}^{BG}(G_{t_{k-1}}, e_{t_{k-1}}) s_{ji}^{BG}(G_{t_{k-1}}, e_{t_{k-1}}) = 1\}.
$$

Since each s_i^{BG} depends on last period's network not taking into account other agents' choices in t_k, T_G will depend on both $G_{t_{k-1}}$ and $e_{t_{k-1}}$.

A similar, more simple updating rule is given by randomly selecting at most one agent who then has the opportunity to adjust his strategy choice given $G_{t_{k-1}}$ similar to [4]. Denote the selected agent by $i^S \in N \cup \emptyset$.

Definition 3. *Under simple unilateral best response dynamics, for all $i \in N$ agent i's strategy is given by*

$$
s_i^{UBR} = \begin{cases} \operatorname{argmax}_{s_i \in \{0;1\}^{n-1}} Eu_i(G_{t_{k-1}}(s_i, s_{-i_{t_{k-1}}})|e_{t_{k-1}}) & \text{if } i = i^S \\ \left(I\left((i,j) \in L_{t_{k-1}}\right)\right)_{j \in \mathcal{N}_{t_{k-1}}^N(i)} & \text{otherwise.} \end{cases}
$$

Again, $I(.)$ denotes an indicator function returning 1 when the argument is true and 0 otherwise. Here the network operator is defined accordingly by

$$
T_G(G_{t_{k-1}}, e_{t_{k-1}}, i_{t_k}^S) = \{G \in \mathcal{G} | (i,j) \in L
$$
$$
\Leftrightarrow s_{ij}^{UBR}(G_{t_{k-1}}, e_{t_{k-1}}, i_{t_k}^S) s_{ji}^{UBR}(G_{t_{k-1}}, e_{t_{k-1}}, i_{t_k}^S) = 1\}.
$$

The agent $i_{t_k}^S$ might for instance be selected by a Poisson process.

Updating of the state

Given a network G_{t_k} and a state vector e_{t_k}, the probabilities α and β and the transmission rules, that is the range of information transmission r, determine $e_{t_{k+1}}$ as follows. Set for the moment $r = 1$. In this case we can calculate the transition probabilities as follows. Denote by $n_{t_k}^e(i) = \sum_{j \in \mathcal{N}_{t_k}^N(i)} e_j^{t_k} - e_i^{t_k}$ the number of direct neighbors of i who are employed at the beginning of period t_k excluding one employed agent.

$$
P_i(e_i^{t_{k+1}} = 1 | e_i^{t_k} = 1) = (1 - \alpha) + \alpha\beta + \alpha(1 - \beta) \times
$$
$$
\left(1 - \prod_{j \in \mathcal{N}_{t_k}^N(i)} \left(1 - \beta \sum_{k=0}^{n_{t_k}^e(j)} \binom{n_{t_k}^e(j)}{k} \frac{(1-\alpha)^{n_{t_k}^e(j)-k} \alpha^k}{k + n_{t_k}^N(j) - n_{t_k}^e(j)}\right)\right),
$$

$$
P_i(e_i^{t_{k+1}} = 1 | e_i^{t_k} = 0) = \beta + (1 - \beta) \times
$$
$$
\left(1 - \prod_{j \in \mathcal{N}_{t_k}^N(i)} \left(1 - \beta \sum_{k=0}^{n_{t_k}^e(j)} \binom{n_{t_k}^e(j)}{k} \frac{(1-\alpha)^{n_{t_k}^e(j)-k} \alpha^k}{k + n_{t_k}^N(j) - n_{t_k}^e(j)}\right)\right).
$$

That is, the employment operator T_e can be defined accordingly by a transition matrix resulting from the conditional probabilities $P_i(.|.)$.

144

Conclusion

We provide a unified formal framework for the analysis of dynamic network formation in a stochastic environment. By separating updating the network links from updating the state of the economy this framework is able to incorporate several solution concepts for network formation models prominent in the literature.

Bibliography

[1] Amann, R. and T. Gall (2006): 'How (not) to Choose Peers in Studying Groups'. *FEEM Working Paper* No. 79.2006.

[2] Bala, V. and S. Goyal (2000): 'A Noncooperative Model of Network Formation'. *Econometrica* 68(5), 1181-1229.

[3] Calvó-Armengol, A. (2004): 'Job Contact Networks'. *Journal of Economic Theory* 114(1), 191-206.

[4] Jackson, M.O. and A. Watts (2002): 'The Evolution of Social and Economic Networks'. *Journal of Economic Theory* 106(1), 265-295.

[5] Jackson, M.O. and A. Wolinsky (1996): 'A Strategic Model of Social and Economic Networks'. *Journal of Economic Theory* 71(1), 44-74.